Solved Problems in Chemistry

Beatriz Elena Soledad Rodríguez

Solved Problems in Chemistry. Volume I.

FIRST EDITION

IBSN: 978-0-557-76441-9

www.lulu.com

Printed in United States of America.

Solved Problems in Chemistry

Beatriz Elena Soledad Rodríguez

Ignorance affirms or denies, science questions.

Voltaire

About the author

Currently works as Associate Professor at the Universidad Católica Andrés Bello in Caracas, Venezuela.

Studied chemistry degree at the Universidad Simón Bolívar (Caracas, Venezuela), subsequently studied a Masters in Chemistry, Specialization in Corporate Finance and Ph.D. in Analytical Sciences (Spain).

She has held positions in private companies in the area of Quality Control, Production, Logistics, General Management and business consulting in the textile, food and pharmaceutical industries.

The student

*T*his book is written to help you study general chemistry.

In each subject area will give a brief theoretical explanation of the concepts to study and then find a series of exercises done.

As you progress through the chapters will encounter increasingly complex problems that will be useful to understand the issues.

Will be presented a number of problems solved, once you understand the explanation, try to solve it without seeing the answer, to see if fully understood.

In addition to the problems solved, will be given a series of exercises for you to practice and find the answer to every problem. It is convenient to solve the exercises in the order given, to help in reasoning.

Table of Contents

MOL AND MOLECULAR FORMULA..**11**

MOL CONCEPT 13

Solved problems..*14*

MOLECULAR MASS CONCEPT   (MOLECULAR MASS) 15

Solved problems..*15*

PERCENTAGE COMPOSITION 17

Solved problems..*18*

EMPIRICAL FORMULA AND MOLECULAR FORMULA 19

Solved problems..*19*

Problems answered *26*

CHEMICAL EQUATIONS..**29**

Solved problems..*31*

STOICHIOMETRY AND STOICHIOMETRIC CALCULATIONS 33

Solved problems..*33*

L IMITING R EAGENT 35

    Solved problems..35

T HEORETICAL YIELD AND P ERCENT YIELD...41

    Solved problems..42

    Problems answered 46

LIQUIDS AND SOLUTIONS **49**

    P ERCENTAGES 52

      Solved problems..53

    M OLARITY..55

      Solved problems..55

    M OLALITY..59

      Solved problems..59

    N ORMALITY 61

      Solved problems..65

      Problems answered 67

TITRATION **69**

    Solved problems..71

    Problems answered 75

PREPARATION OF SOLUTIONS...**77**

    Solved problems..80

    Problems answered 83

STOICHIOMETRY OF MIXTURES **85**

PERCENT PURITY 97

    Solved problems...*88*

    Problems answered *98*

REDOX REACTIONS **99**

BALANCING REDOX REACTIONS 101

    Solved problems...*104*

REDOX TITRATIONS 117

    Solved problems...*118*

    Problems answered *124*

ATOMIC MASS OF THE ELEMENTS **128**

MOL AND MOLECULAR FORMULA

What we know is a drop of water, what we ignore is the ocean.

Isaac Newton.

Mol concept

A mol is the amount of substance that contains as many elementary entities (such as atoms, molecules or other particles) as there are atoms in exactly 12 grams of carbon-12 isotope.

$$1\ mol = 6.022x10^{23}\ particles$$

As the molar mass of Carbon-12 is 12 grams, then 1 mol of Carbon-12 weighs 12 grams

$$N^\circ\ moles = \frac{g}{MM}$$

Where:

$N^\circ\ moles$ = Number of moles

g = grams

MM = Molecular Mass or Molecular Weight (MW)

Solved problems

a) **Calculate the number of moles found in 5.84 grams of potassium**

Resolution:

Atomic Mass of potassium (chemical symbol K) is 39,098 g / mol.

Using equation (1), we have:

$$N^\circ\ moles = \frac{5.84\,grams}{39.098\,grams\ /\ mol} = 0.1494\,moles$$

Answer: 0.1494 moles.

b) **Calculate the number of atoms of Aluminium (Al), found in 14.6 grams of Al.**

Resolution:

Atomic Mass of Al is 26.982 grams/mol.

$$14.6\,grams\ of\ Al \times \frac{1mol\ Al}{26.982\,grams} \times \frac{6.022x10^{23}\ atoms}{1mol} = 3.26\,x10^{23}\ atoms\ of\ Al$$

Answer: 3.26×10^{23} atoms of Al

Molecular Mass Concept
(Molecular Mass)

Molecular Mass or Molecular Weight is the sum of the masses of the atoms that constitute the molecule.

Solved problems

a) Determine the Molecular Mass of calcium hydroxide.

Molecular Mass: $Ca(OH)_2$

Resolution:

The calcium hydroxide molecule contains:

1 atom of Calcium, 2 oxygen atoms and 2 hydrogen atoms.

Then:

1 atom of calcium x 40,078 g / mol =	40,078 g / mol
2 atoms of oxygen x 15,999 g / mol =	31,998 g / mol
2 Hydrogen atoms x 1.0079 g / mol =	2.0158 g / mol
	74.0918 g/mol

Answer: 74.0918 g/mol

b) Determine the Molecular Mass of copper sulfate penta-hydrate:
Molecular Formula: $CuSO_4$ x 5 H_2O.

Resolution:

The molecule of copper sulfate penta-hydrate contains:

1 atom of copper, 1 atom of sulfur, oxygen atoms 9 and 10 hydrogen atoms.

Then:

1 atom of Copper x 40.078 g/mol =	40.078 g/mol
1 atom of Sulfur x 32.066 g/mol =	32.066 g/mol
9 atoms of Oxygen x 15.999 g/mol =	143.991 g/mol
10 atoms of hydrogen x 1.0079 grams/mol =	<u>10.079 g/mol</u>
	226.214 g/mol

Answer: 226.214 g/mol

Percentage composition

Knowing the formula of a compound, its chemical composition can be expressed as the percentage of mass of each element in the compound.

$$Percentage\ composition\ of\ an\ element = \frac{n \times AM\ element}{MM\ compound} \times 100$$

Where:

n = Atoms of element in compound

AM = Atomic Mass of element

MM = Molecular Mass of compound

a) Determine the percentage composition of the phosphoric acid molecule.

Resolution:

Molecular Formula: H_3PO_4 MM = 97.99 grams/mol

For Hydrogen: $\% \, H = \dfrac{3 \times 1.008}{97.99} \times 100 = 3.086 \quad \% \, H$

For Phosphor: $\% \, P = \dfrac{1 \times 30.97}{97.99} \times 100 = 31.61 \quad \% \, P$

For Oxygen: $\% \, O = \dfrac{4 \times 15.99}{97.99} \times 100 = 65.31 \quad \% \, O$

Answer: H: 3.086 %;   P 31.61 %;   O: 65.31 %

b) Determine the percentage composition of Ethanol.

Resolution:

Molecular Formula: CH_3CH_2OH MM = 46.047 grams/mol

For Carbon: $\% \, C = \dfrac{2 \times 12}{46.047} \times 100 = 52.12 \quad \% \, C$

For Hydrogen: $\% \, H = \dfrac{6 \times 1.008}{46.047} \times 100 = 13.13 \quad \% \, H$

For Oxygen: $\% \, O = \dfrac{1 \times 15.99}{46.047} \times 100 = 34.73 \quad \% \, O$

Answer: H: 13.13 %;   C: 52.12 %;   O: 34.73 %

Empirical Formula and Molecular Formula

Empirical Formula (EF): It is the simplest formula of a molecule

Molecular Formula (MF): can be equal to or an integral multiple of the Empirical Formula.

e.g.:

For Hydrogen peroxide, EF = HO and MF= H_2O_2

For benzene, EF = CH and MF is C_6H_6

Solved problems

a) The analysis of a sample reveals a pure compound that contains carbon 27.27% and 72.72% Oxygen. What is the empirical formula?.

Resolution:

AM Carbon = 12 grams/mol   AM Oxygen = 15.999 grams/mol

N° moles of atoms of C = $27.27 \ g \ C \times \dfrac{1 \ mol \ C}{12 \ g \ C} = 2.2725 moles$

Nº moles of atoms of O = $72.72 g O \times \dfrac{1 \ mol \ O}{15.999 \ g \ O} = 4.5453 \ moles$

It would be written: $C_{2.2725} O_{4.5453}$; but chemical formulas must be written with integers, so we must divide by the smallest number:

$$C\frac{2.2725}{2.2725}O\frac{4.5453}{2.2725} = CO_2 \qquad \text{Empirical Formula is: } CO_2$$

b) It was determined that a sample of a compound with mass equal to 25.71 grams, containing 7.14 grams of magnesium, 6.03 grams of phosphorus and 12.54 grams of Oxygen. Determine the percentage composition and empirical formula.

Resolution:

AM Mg= 24.305 grams/mol,   AM P= 30.974 grams/mol,   AM O= 15.999 grams/mol

For determinate the percentage composition

$$\% \, Mg = \frac{grams \; of \; Mg \; in \;\;\; compound}{grams \; of \; compound} \times 100 = \frac{7.14}{25.71} \times 100 = 27.77 \, \%$$

$$\% \, P = \frac{6.03 \; grams \; P}{25.71 \; grams \; of \; compound} \times 100 = 23.45 \, \%$$

$$\% \, O = \frac{12.54 \; grams \; O}{25.71 \; grams \; of \; compound} \times 100 = 48.77 \, \%$$

Percentage composition: % Mg = 27.77 %, % P = 23.45 %, % O = 48.77 %.

For determinate the Empirical Formula:

$$\text{N° moles of atoms of Mg} = 27.27 g \times \frac{1 \; mol \; Mg}{24.305 g} = 1.14 \; moles$$

$$\text{N° moles of atoms of P} = 23.45 \; g \times \frac{1 \; mol \; P}{30.974 \; g} = 0.76 \; moles$$

N° moles of atoms of O = $48.73 \; g \times \dfrac{1 \; mol \; O}{15.999 \; g} = 3.05 \; moles$

Now we have: $Mg_{1.14} \; P_{0.76} \; O_{3.05}$, but it must be written with integers, so we must divide by the smallest number:

$Mg\dfrac{1.14}{0.76} P\dfrac{0.76}{0.76} O\dfrac{3.05}{0.76} = Mg_{1.5} \; P_1 \; O_4$ But since it can give a fractional number is multiplied by two, with:

$Mg_3P_2O_8$, which corresponds to the following formula: $Mg_3(PO_4)_2$

To know the Molecular Formula of a compound, we must know its empirical formula and its molecular weight.

$$Factor = \dfrac{Molecular \; Formula \; weight}{Empirical \; Formula \; weight}$$

c) A sample of 15 grams of an organic compound containing only Carbon, Hydrogen and Oxygen, was placed on a train and only combustion products are obtained as 32.96 grams of CO2 and 6 grams of H2O. If the molecular weight of the compound is 180 grams / mol, determine its Empirical Formula and Molecular Formula.

Resolution:

First it is necessary to determine the grams of Carbon, Hydrogen and Oxygen found in the compound. The Carbon and Hydrogen is obtained from the amounts derived from CO_2 and H_2O and Oxygen is obtained by difference.

To calculate the grams of Carbon:

$$32.96 \; g \; of \; CO_2 \times \frac{12 \; g \; of \; C}{44 \; g \; of \; CO_2} = 8.989 \; g \; C$$

To calculate the grams of Hydrogen:

$$6 \; g \; of \; H_2O \times \frac{2 \; g \; of \; H}{18 \; g \; of \; H_2O} = 0.66 \; g \; of \; H$$

To calculate the grams of Oxygen:

$$g \; of \; O = g \; of \; Compound - g \; of \; C - g \; of \; H$$

$$g \; of \; O = 15 \; g \; Compound - 8.989 \; g \; C - 0.66 \; g \; H = 5.351 \; g \; of \; O$$

Now calculate the percentage composition of each element in the compound:

$$\% \; of \; C = \frac{8.989 \; g \; of \; C}{15 \; g \; of \; Compound} \times 100 = 59.92 \; \%$$

$$\% \ of \ H = \frac{0.66 \ g \ of \ H}{15 \ g \ of \ Compound} \times 100 = 4.4 \ \%$$

$$\% \ of \ O = \frac{5.351 \ g \ of \ O}{15 \ g \ of \ Compound} \times 100 = 35.67 \ \%$$

Percentage composition: 59.92 % C, 4.4 % H and 35.67 % O

To calculate the Empirical Formula:

$$N° moles \ of \ atoms \ of \ C = 59.92 \ g \ of \ C \times \frac{1 \ mol \ of \ C}{12 \ g} = 4.99 \quad moles$$

$$N° moles \ of \ atoms \ of \ H = 4.4 \ g \ of \ H \times \frac{1 \ mol \ of \ H}{1.008 \ g} = 4.4 \ moles$$

$$N° moles \ of \ atoms \ of \ O = 35.67 \ g \ of \ O \times \frac{1 \ mol \ of \ O}{15.999 \ g} = 2.22 \ moles$$

We have: $C_{4.99} \ H_{4.4} \ O_{\ 2.22}$ Is now divided by the smaller number: $C\frac{4.99}{2.22}H\frac{4.4}{2.22}O\frac{2.22}{2.22}$ being $C_{2.25}H_2O_1$, but we still have a fractional number, then the coefficients are multiplied by a number to obtain an integer, in this case is multiplied by 4 and is obtained: $C_9H_8O_4$

The weight of Empirical Formula is: 180 grams/mol and as the weight of Molecular Formula is 180 grams/mol, then $MF = MF$

d) It is burned in a combustion train 30.5 grams of an organic compound composed of C, H and N, to give 88 grams of CO2, 11.5 grams of NO2 and 27 grams of H2O. Determine the percentage composition of the compound and Molecular Formula.

Resolution:

To calculate grams of Carbon:

$$88 \ g \ CO_2 \times \frac{12 \ g \ C}{44 \ g \ CO_2} = 24 \ g \ C$$

To calculate grams of Hydrogen:

$$27 \ g \ H_2O \times \frac{2 \ g \ H}{18 \ g \ H_2O} = 3 \ g \ H$$

To calculate grams of Nitrogen:

$$11.5 \ g \ NO_2 \times \frac{14 \ g \ N}{46 \ g \ NO_2} = 3.5 \ g \ N$$

To calculate the percentage composition of the compound:

$$\% \ of \ C = \frac{24 \ g \ of \ C}{30.5 \ g \ of \ Compound} \times 100 = 78.69 \ \% \ C$$

$$\% \ of \ H = \frac{3 \ g \ of \ H}{30.5 \ g \ of \ Compound} \times 100 = 9.84 \ \% \ H$$

$$\% \ of \ N = \frac{3.5 \ g \ of \ N}{30.5 \ g \ of \ Compound} \times 100 = 11.48 \ \% \ N$$

Percentage composition: 78.69 % C, 9.84 % H y 11.48 % N.

To determine the Empirical Formula

$$N° \text{ moles of atoms of } C = 78.69 \text{ g of } C \times \frac{1 \text{ mol of } C}{12 \text{ g}} = 6.56 \quad moles$$

$$N° \text{ moles of atoms of } H = 9.84 \text{ g of } H \times \frac{1 \text{ mol of } H}{1.008 \text{ g}} = 9.76 \text{ moles}$$

$$N° \text{ moles of atoms of } N = 11.48 \text{ g of } N \times \frac{1 \text{ mol of } N}{14.007 \text{ g}} = 0.82 \text{moles}$$

We have $C_{6.56} H_{9.76} N_{0.82}$ divided by the lowest value: $C\frac{6.56}{0.82} H\frac{9.76}{0.82} N\frac{0.82}{0.82}$
and must be the EF: $C_8H_{12}N$, the weight is: 122 grams/mol. As the weight of EF is the same of MW, then *EF=MF*.

Problems answered

1.- Calculate the number of moles found in 45.62 grams of nickel.

ANSWER. 0.777 moles

2.- Calculate the number of atoms found in 85.94 grams of Chromo.

ANSWER. 9.95 x 10^{23} atoms

3.- Determine the Molecular Weight of H_2SO_4.

ANSWER. 98 g/mol

4.- Determine the Molecular Weight of Na_3PO_4.

ANSWER. 163.97 g/mol

5.- Determine the Molecular Weight of $Pb(NO_3)_2$.

ANSWER. 331.2 g/mol

6.- Determine the Molecular Weight of $(NH_4)_2CO_3$.

ANSWER. 96.09 g/mol

7.- Determine the percentage composition of Sulfur Dioxide (SO_2).

ANSWER. S = 50 %, O= 50 %

8.- Determine the percentage composition of sodium thiosulfate ($Na_2S_2O_3$).

ANSWER. Na = 29,11 %, S =40,51 %, 30,38 %

9.- Determine the percentage composition of Benzene (C_6H_6).

ANSWER. C = 93,31 %, H = 7,69 %

10.- Determine the percentage composition of $CaCO_3$.

ANSWER. Ca = 40 %, C = 12 %, O = 48 %

CHEMICAL EQUATIONS

It is better to know something of everything, know everything about one thing.

Blaise Pascal.

Chemical Equations: describes the chemical changes in a reaction in which one or more substances are changed into one or more products.

All substances must be represented by means of the formulas that describe the number of atoms of each type on each side of the equation must be equal to that equation is *balanced*.

Solved problems

a) In the combustion of glucose ($C_6H_{12}O_6$) is produced CO_2 and H_2O. Write the chemical equation and balance it.

Resolution:

Chemical equation is:

$$C_6H_{12}O_6 \;+\; O_2 \;\rightarrow\; CO_2 \;+\; H_2O$$

REAGENTS   PRODUCTS

The number of Carbon atoms in the reactant side must equal the product side, so you should put a 6 in the CO_2, so does the Hydrogen, so place a 6 in H_2O, and to make it balanced the Oxygen is put a 6 in the O_2, leaving the equation as:

$$C_6H_{12}O_6 \;+\; 6\,O_2 \;\rightarrow\; 6\,CO_2 \;+\; 6\,H_2O$$

b) Balance the following equation:

$$Pb(NO_3)_2 \text{ (ac)} + NaCl \text{ (ac)} \rightarrow PbCl_2 \text{ (s)} + NaNO_3 \text{ (ac)}$$

Resolution:

For the number of each atom present in the reactant side is equal to the number of each atom present in the products is equal, must be balanced as follows:

$$Pb(NO_3)_2 \text{ (ac)} + 2 \, NaCl \text{ (ac)} \rightarrow PbCl_2 \text{ (s)} + 2 \, NaNO_3 \text{ (ac)}$$

Stoichiometry and Stoichiometric Calculations

From chemical equations can calculate the quantities of various substances involved in chemical reactions

Solved problems

a) In the combustion of methane, say how many Oxygen molecules are required to react with 25 molecules CH_4.

Resolution:

The balanced equation is: $\qquad CH_4 + 2 O_2 \rightarrow CO_2 + 2 H_2O$

The stoichiometry indicates that one molecule of methane reacts with two molecules of Oxygen, therefore we have:

$$N°\,molecules\ of\ O_2 = 25\ moleculess\ of\ CH_4 \times \frac{2\ molecules\ of\ O_2}{1\ molecules\ of\ CH_4} = 50\ molecules\ O_2$$

Answer: It takes then 50 CH_4 molecules to react with 25 molecules of O_2.

b) 20 grams of calcium is reacted with water. Say how many moles of water are required to react completely with calcium.

Resolution:

The balanced equation is:

$$Ca + 2\ H_2O \rightarrow Ca(OH)_2 + H_2(g)$$

$$N°\ moles\ of\ Ca = \frac{20\ g}{40\ g/mol} = 0{,}5\ moles\ of\ Ca$$

By stoichiometry, 1 mol of calcium reacts with 2 moles of water, so we have:

$$N°molecules\ of\ H_2O = 0{,}5\ moles\ of\ Ca \times \frac{2\ moles\ of\ H_2O}{1\ mol\ of\ Ca} = 1\ mol\ of\ H_2O$$

Answer: 1 mol of water.

Limiting Reagent

In a chemical reaction, most of the time there is a reagent that is completely consumed and there are others that are in excess, which is completely consumed is the limiting reagent

Solved problems

a) 47.65 grams of magnesium chloride is reacted with 42.08 grams of potassium hydroxide, according to the reaction:

$$MgCl_2 + 2\,KOH \rightarrow Mg(OH)_2 + 2\,KCl$$

Determine:

1.- Limiting Reagent

2.- Grams of excess reagent

3.- Grams produced of potassium chloride.

Resolution:

First calculate the number of moles of each of the items involved in the reaction:

$$N° \ moles \ MgCl_2 = \frac{47,65 \ g}{95,3 \ g/mol} = 0,5 \ moles$$

$$N°moles \ KOH = \frac{42,08 \ g}{56,1 \ g/mol} = 0,75 \ moles$$

The stoichiometric ratio tells us that 1 mol of magnesium chloride reacts with 2 moles of potassium hydroxide; therefore we must review how many moles of KOH react 0.5 moles of magnesium chloride.

$$N° \ moles \ KOH \ react = 0,5 \ moles \ of \ MgCl_2 \times \frac{2 \ mol \ KOH}{1 \ mol \ MgCl_2} = 1 \ mol \ of \ KOH$$

But we only have 0.75 mol of KOH, so we have to calculate how many moles of $MgCl_2$ react 0.75 moles of KOH.

$$N° \ moles \ MgCl_2 \ react = 0,75 \ moles \ of \ KOH \times \frac{1 \ mol \ MgCl_2}{2 \ moles \ KOH} = 0,375 \ moles$$

We must to evaluate the following:

$$MgCl_2 + 2 \ KOH \rightarrow Mg(OH)_2 + 2 \ KCl$$

	MgCl₂	KOH	Mg(OH)₂	KCl
Initial moles	0,5	0,75	0	0
Moles that react	0,375	0,75	0	0
Final moles	0,125	0	0,375	0,75

KOH is the reagent completely consumed, therefore it is the limiting reactant (LR), and the excess reagent is $MgCl_2$, so called excess reactant (ER).

1.- Limiting reagent: KOH

Using the stoichiometric ratio, we can calculate the moles formed both magnesium hydroxide and potassium chloride.

$$N° \text{ moles } Mg(OH)_2 \text{ formed} = 0,75 \text{ moles } KOH \times \frac{1 \, mol \, Mg(OH)_2}{2 \, moles \, KOH} = 0,375 \, moles$$

$$N° moles \, KCl \text{ formed} = 0,75 \text{ moles } KOH \times \frac{2 \, moles \, KCl}{2 \, moles \, KOH} = 0,75 \, moles$$

To calculate the grams of excess reactant ($MgCl_2$), we use the final moles after the reaction, in this case 0.125 moles.

$$g \, MgCl_2 = N° moles \times PM = 0,125 \, moles \times 95,3 \, g/mol = 11,91 \, grams$$

2.- Grams of excess reagent ($MgCl_2$): 11,91 g

Using KCl formed moles calculated the grams produced:

$$g \, KCl = N° moles \times PM = 0,75 \, moles \, KCl \times 74,6 \, g/mol = 55,95 \, g$$

3.- Grams produced in KCl: 55,95 g

b) Calculate the greatest amount of HCN that can be obtained when reacted 29.4 g of sodium cyanide with sulfuric acid 49 g, according to the reaction:

$$2 \, NaCN \; + \; H_2SO_4 \rightarrow Na_2SO_4 \; + \; 2 \, HCN$$

Resolution:

First we calculate the number of moles of each reagent:

$$N° \; moles \; NaCN = \frac{29,4g}{49 \; g/mol} = 0,6 \; moles$$

$$N° moles \; H_2SO_4 = \frac{49g}{98 \; g/mol} = 0,5 \; moles$$

The stoichiometric ratio tells us that 2 moles of NaCN react with 1 mole of H_2SO_4, we now need to see how many moles of sulfuric acid reacts 0.6 mol of sodium cyanide.

$$N°moles \; de \; NaCN \; that \; react = 0,5 \; moles \; of \; H_2SO_4 \times \frac{2 moles \; NaCN}{1mol \; H_2SO_4} = 1 \; mol \; NaCN$$

We don't have enough NaCN to react with all sulfuric acid, so it is necessary to calculate the number of moles of sulfuric acid reacting with sodium cyanide:

$$N°moles \; H_2SO_4 \; that \; react = 0,6 \; moles \; NaCN \times \frac{1 \, mol \; H_2SO_4}{2 \; moles \; NaCN} = 0,3 \; moles \; of \; H_2SO_4$$

We see that the limiting reagent is the NaCN and H_2SO_4 reagent in excess.

Evaluating the stoichiometry of the reaction and the moles involved are:

$$2\,NaCN \quad + \quad H_2SO_4 \rightarrow Na_2SO_4 \quad + \quad 2\,HCN$$

Initial moles	0,6	0,5	0	0
Moles that react	0,6	0,3	0	0
Final moles	0	0,2	0,3	0,6

The maximum number of moles of HCN that can be obtained is 0.6, so if you want to know the grams produced do the following operation:

$$g\ HCN = 0{,}6\ moles\ HCN \times 27\,g\,/\,mol = 12{,}6\ grams\ of\ HCN$$

Theoretical yield and Percent yield

Theoretical yield: When chemical reaction occurs, if you spend the entire limiting reagent, the amount of product to be formed is the theoretical amount to be obtained. This is called the Theoretical yield.

Actual yield: the amount of product produced in a chemical reaction

Percent yield: Current yield relates to the theoretical yield using the expression

$$\% \ yield = \frac{actual \ yield}{theoretical \ yield} \times 100$$

a) A solution containing 11.7 g of aluminum hydroxide reacted with 6.3 g of nitric acid by reaction:

$$Al(OH)_3 \quad + \quad 3\,HNO_3 \quad \rightarrow \quad Al(NO_3)_3 \quad + \quad 3\,H_2O$$

If you have obtained 6.5 grams of aluminum nitrate, determine the Percent yield of the reaction.

Resolution:

$$N° \; moles \; Al(OH)_3 = \frac{11,7\;g}{77,98\;g/mol} = 0,15\;moles$$

$$N° \; moles \; HNO_3 = \frac{6,3\;g}{63\;g/mol} = 0,1\;mol$$

By stoichiometry, 1 mol of aluminum hydroxide reacts with 3 moles of nitric acid, to calculate the limiting reagent:

$$N° \; moles \; HNO_3 \; reaccionan = 0,15 \; moles \; Al(OH)_3 \; \frac{3\;moles\;HNO_3}{1\;mol\;Al(OH)_3} = 0,45\;moles$$

But we only have 0.1 moles of HNO_3, so we must do the calculation with the number of moles of aluminum hydroxide react:

$$N° \; moles \; Al(OH)_3 \, react = 0,1 \; mol \; HNO_3 \times \frac{1\;mol\;Al(OH)_3}{3\;moles\;HNO_3} = 0,033 \; moles \; Al(OH)_3$$

We see that the limiting reagent is nitric acid. Applying the stoichiometry we have:

$$Al(OH)_3 \quad + \quad 3\,HNO_3 \quad \rightarrow \quad Al(NO_3)_3 \quad + \quad 3\,H_2O$$

	$Al(OH)_3$	$3\,HNO_3$	$Al(NO_3)_3$	$3\,H_2O$
Initial moles	0,150	0,1	0	0
Moles that react	0,033	0,1	0	0
Final moles	0,0117	0	0,033	0,1

Theoretical grams $Al(NO_3)_3 = 0,033 moles\ Al(NO_3)_3 \times 213\ g/mol = 7,03 grams$

Grams obtained= 6,5 g.

$$\% \ yield = \frac{6,5\ g\ obtained}{7,03\ gramos\ theoretical} \times 100 = 92,46\ \%$$

b) **A solution X containing 40.4 grams of calcium phosphate, is mixed with another solution and sodium sulfate by the reaction.**

$$Ca_3(PO_4)_2 + 3\,Na_2SO_4 \rightarrow 3\,CaSO_4 + 2\,Na_3PO_4$$

It is known that both reactants are fully consumed. If the percentage of yield obtained sodium phosphate was 87.5%, calculate:

1. Grams of sodium phosphate obtained.

2. Grams of sodium sulfate present in the solution Y

Resolution:

The first thing to calculate is the number of moles of calcium phosphate:

$$N° \, moles \, Ca_3(PO_4)_2 = \frac{40.4 \; grams}{202 \; grams \, / \, mol} = 0.2 \; moles$$

Using the stoichiometry of the reaction is:

$$Ca_3(PO_4)_2 + 3\,Na_2SO_4 \rightarrow 3\,CaSO_4 + 2\,Na_3PO_4$$

	$Ca_3(PO_4)_2$ +	$3\,Na_2SO_4$ →	$3\,CaSO_4$ +	$2\,Na_3PO_4$
Initial moles	0,2	y	0	0
Moles that react	0.2	y	0	0
Final moles	0	0	a	b

The maximum number of moles of sodium phosphate can be obtained is:

$$N° moles \, Na_3PO_4 = 0.2 \; moles \, Ca_3(PO_4)_2 \times \frac{2 \; moles \; Na_3PO_4}{1 \; mol \; Ca_3(PO_4)_2} = 0.4 \; moles$$

The maximum number of grams of sodium phosphate can be obtained is:

$$grams\ Na_3PO_4 = 0.4\ moles \times 164\ g/mol = 65.5\ grams$$

For the percentage yield equation, we have:

$$Grams\ Na_3PO_4 = \frac{87.5\%\ ren\ dim\ iento}{100} \times 65.5\ grams\ theorethical = 57.3\ grams$$

1.- $Grams\ Na_3PO_4 = 57.3\ grams$

To calculate the grams of sodium sulfate present in the solution Y, again working with the stoichiometry of the reaction:

$$N°\ moles\ Na_2SO_4 = 0.2 moles Ca_3(PO_4)_2 \times \frac{3\ moles\ Na_2SO_4}{1\ mol\ Ca_3(PO_4)_2} = 0.6\ moles\ Na_2SO_4$$

To calculate the grams of sodium sulfate present in the solution Y, applies the equation:

$$grams\ Na_2SO_4 = 0.6\ moles \times 142\ grams/mol = 85.2\ grams$$

2.- Grams of sodium sulfate present in the solution Y are: 85.2 grams.

1.- Balance the following reactions:

$$H_2SO4 + Al(OH)_3 \rightarrow Al_2(SO_4)_3 + H_2O$$

$$Fe(OH)_3 + HMnO_4 \rightarrow Fe(MnO_4)_3 + H_2O$$

$$CaF_2 + H_2SO_4 \rightarrow CaSO_4 + HF$$

ANSWER. $3\,H_2SO_4 + 2\,Al(OH)_3 \rightarrow Al_2(SO4)_3 + 6\,H_2O$

$$Fe(OH)_3 + 3HMnO_4 \rightarrow Fe(MnO_4)_3 + 3\,H_2O$$

$$CaF_2 + H_2SO_4 \rightarrow CaSO_4 + 2HF$$

2. 25 grams of FeS is reacted with 40 grams of O_2. according to the reaction:

$$FeS + O_2 \rightarrow Fe_2O_3 + SO_2$$

Balance the reaction and determine: a) Limiting reagent. b) Excess reagent c) The grams of ferric oxide produced.

ANSWER. a) FeS, b) O_2, c) 22,72 g.

3. When 30 grams of $KClO_3$ decompose by the reaction given below:

$$KClO_3 \rightarrow KCl + O_2$$

How many grams of KCl and O_2 are produced?

ANSWER. 18,26 grams KCl and 11,75 grams O_2

4. If you have 49 grams of sulfuric acid and these react completely with all the tin IV oxide present:

$$SnO_2 + H_2SO_4 \rightarrow Sn(SO_4)_2 + H_2O$$

How many grams of tin sulfate IV are produced?

The moles of tin oxide reacted IV.

ANSWER. a) 77,7 g   b) 37,68 g

5. A solution containing two moles of sulfuric acid is reacted with another solution containing 4 moles of calcium phosphate, according to the equation:

$$H_2SO_4 + Ca_3(PO_4)_2 \rightarrow CaSO_4 + H_3PO_4$$

If the current yield of the reaction was 85%, determine the grams of calcium sulfate and phosphoric acid obtained.

ANSWER. 77.07 grams of calcium sulfate and 37.03 grams of phosphoric acid

6. In the following reaction:

$$H_3PO_4 + Pt(OH)_4 \rightarrow Pt_3(PO_4)_4 + H_2O$$

If you get 194.4 grams of water at a yield of 90%, determine the grams of phosphoric acid and platinum hydroxide used for the reaction.

ANSWER. 392 grams of phosphoric acid, and 789.3 grams of platinum hydroxide.

LIQUIDS AND SOLUTIONS

In the end, scientists are lucky people: we play what we want for life.

Lee Smolin.

This chapter is necessary to disclose the terms of solution and dissolution.

S olution: can be defined as a solution to a homogeneous mixture of pure substances in which without precipitation.

S olvent: is the medium in which solutes dissolve

T he solutions are classified as:

Saturated solutions:

Are those in which the dissolution of the solute and once equilibrium is reached, no more solid can dissolve without causing the crystallization of an equal mass of dissolved ions.

Unsaturated solutions:

Are those that can dissolve even greater amount of solute.

Supersaturated solutions:

Metastable solutions are prepared at high temperatures and cooled slowly without stirring. If this solution is disturbed, or introduce dust particles or crystals (seeds), crystallize and become solid in a saturated solution.

To work with the solutions you need to know how much solute is dissolved in the solvent, which is known as concentration. Let's look at some ways of expressing these units.

Percentages

Weight percentage:

Is the percentage ratio between the mass of solute and mass of the solution, and is given by the formula:

$$\% \ solute = \frac{solute \ mass}{solution \ mass} \times 100$$

Weight-volume percentage:

Is the percentage ratio between the mass of solute and volume of the solution, and is given by

$$\% \ solute = \frac{solute \ mass}{solution \ volume} \times 100$$

Volume-volume percentage:

Is the percentage ratio between the volume of solute and volume of the solution, and is given by

$$\% \ v/v = \frac{volume \ of \ solute}{volume \ of \ solution} \times 100$$

a) Calculate the mass of copper sulfate (CuSO₄), containing 300 grams of a copper sulfate solution at 5%.

Resolution:

We have:

$\% \ solute = 5\%$
$grams \ of \ solution = 300g$
$grams \ of \ solute = ?$

Substituting in the above equation

$$solute \ mass = \frac{\% \ solute \times solution \ mass}{100} = \frac{5\% \times 300 \ g}{100} = 15 \ g \ of \ CuSO_4$$

Answer: 15 grams of copper sulfate

b) Calculate the mass of a copper sulfate solution at 5% containing 25 grams of copper sulfate.

Resolution:

Substituting in the above equation

$$solution \ mass = \frac{solute \ mass \times 100}{\% \ solute} = \frac{25 \ grams \times 100}{5\%} = 500 \ g \ solution$$

Answer: 500 grams.

c) Calculate the mass of copper sulfate $CuSO_4$ in 150 mL of copper sulfate solution at 5%, whose density is 1.05 g / mL at a temperature of 25 ° C.

Resolution:

To determine the mass of the solution must work with the density formula

$$d = \frac{mass}{volume}$$

$mass\ of\ solution = density \times volume = 1.05\ g/ml \times 150\ ml = 157.5g\ de\ solution$

$g\ CuSO_4 = \dfrac{g\ of\ solution \times \%\ solute}{100} = \dfrac{157.5\ g\ of\ solution \times 5\ \%}{100} = 7.88\ g\ of\ CuSO_4$

Answer: 7.88 grams of copper sulfate.

d) What volume of a solution of copper sulfate to 10% contains 15 g of CuSO4, when the density of the solution is 1.10 g / mL at a temperature of 25 ° C?

Resolution:

$$solution\ mass = \frac{g\ solute \times 100}{\%\ solute} = \frac{15\ g\ CuSO_4 \times 100}{10\%} = 150\ g\ solution$$

Using the same formula of density:

$$volume\ of\ solution = \frac{mass}{density} = \frac{150\ g}{1.10\ g/ml} = 136.36\ ml$$

Answer: 136.36 mL

Molarity

Molarity (**M**) is defined as the number of moles of solute per liter of solution

$$M = \frac{n^{\circ}\,moles\;solute}{Volume\;of\;solution}$$

Solved problems

a) Calculate the Molarity of a copper sulfate solution prepared by dissolving 4.8 g of copper sulfate in 2.00 liters of solution.

Resolution:

To calculate the number of moles, it is necessary to know the Molecular Weight of $CuSO_4$.

MW $CuSO_4 = 159.5$ g/mol

$$N^{\circ}\,moles = \frac{g}{PM} = \frac{4.8\;g\;CuSO_4}{159.5\;g\,/\,mol} = 0.030\;moles\;CuSO_4$$

Then apply the equation relating the number of moles in the volume to know the Molarity:

$$M = \frac{n^{\circ}\,moles}{V} = \frac{0.03\;moles\;CuSO_4}{2.00\;liter} = 0.015\;moles\,/\,liter$$

$$M\ CuSO_4 = 0.015\ moles/liter$$

b) Calculate the mass of Mg(OH)₂ needed to prepare 300 mL of a 0.5 Molar solution of magnesium hydroxide.

Resolution:

The MW of magnesium hydroxide is: 58.31 g / mol.

Using the Molarity equation , we have:

$$N°\,moles = M \times V = 0.5\ moles\,/\,L \times 0.3\ L = 0.15\ moles$$

We know:

$$N°\,moles = \frac{g}{MW} \Rightarrow g = N°\,moles \times MW = 0.15\ moles \times 58.31\ g\,/\,mol = 8.75\ grams$$

Grams of magnesium hydroxide = 8.75 g

c) Calculate the Molarity of sulfuric acid container whose label is reported that the% by mass is 95.8% and its density is 1.80 g / mL.

Resolution

MW of H_2SO_4 = 98,1 g/mol

Using the formula for density, we know how much it weighs 1 liter of solution:

$$d = \frac{m}{V} \Rightarrow \quad m = d \times V = 1.80\ g\,/\,mL \times 1000\ mL = 1800\ g$$

As the % by mass is 95.8%, using the formula of % by mass

$$\% \ soluto = \frac{solute \ mass}{mass \ of \ solution} \times 100 \Rightarrow \ m \ solute = \frac{\% \ solute \times mass \ of \ solution}{100}$$

$$m \ solute = \frac{95.8 \ \% \times 1800g}{100} = 1724.4 \ g \ of \ H_2SO_4$$

As these 1724.4 g of sulfuric acid are present in 1 liter of solution, then we can determine the number of moles present in the volume of solution, giving us the Molarity of the solution

$$N^\circ \ moles = \frac{g}{PM} = \frac{1724.4 \ g}{98.1 \ g / mol} = 17.578 \ moles$$

$$M = \frac{N^\circ \ moles}{V} = \frac{17.578 \ moles}{1 \ L} = 17.578 \ moles / L$$

Molarity of sulfuric acid solution is 17,578 moles / L

Molality

Molality (**m**) is defined as the number of moles of solute per kilogram of solvent

$$m = \frac{N° \, mole \, s \, solute}{Kg \; solvent}$$

Solved problems

a) What is the molality of a solution that contains 4.5 grams of glucose in 40 grams of water?.

Resolution:

First we must know the MW of glucose. Its Molecular Formula is $C_6H_{12}O_6$, so it's MW = 180 g / mol

Now we calculate the No. of moles:

$$N° \, moles = \frac{4.5 \; grams}{180 \; g/mol} = 0.025 \; moles$$

Using the formula for molality, and knowing that the body of water is 0.040 Kilograms we have:

$$m = \frac{0.025 \; moles}{0.04 \; Kg} = 0.625 \; molal$$

The glucose solution is 0.625 molal

c) How many grams of water should be used to dissolve 54 grams of glucose, if you wish to prepare a 1.25 molal solution?.

Resolution:

We have to know the number of moles of glucose present in the solution

$$N^\circ moles = \frac{g}{MM} = \frac{54\ g}{180\ g/mol} = 0.3\ moles\ of\ glu\cos e$$

And with the number of moles calculated the grams of water

$$Kg\ water = \frac{N^\circ moles\ of\ glu\cos e}{molality} = \frac{0.3\ moles}{1.25\ molal} = 0.24\ Kg\ water$$

Grams of water required: 240 grams

Normality

Normality (**N**) is defined as the number of equivalents per liter of solution.

$$N = \frac{N° \, equivalents}{Volume \; of \; solution}$$

Number of equivalents is defined as:

$$N° \, equivalents = \frac{grams \; of \; solute}{Equivalent \; Weight}$$

However, it is necessary to define the concept of Equivalent Weight (EW)

Equivalent Weight: in the acid-base reactions, the equivalent weight is defined as the mass of acid (expressed in grams) that provides $6.022 * 10^{23}$ Hydrogen ions (1 mol) or that react with $6.022 * 10^{23}$ (1 mol) of hydroxide ions.

A.- EQUIVALENT WEIGHT OF AN ACID:

Monoprotics acids.

$$HCl \quad \xrightarrow{H_2O} \quad H^+ \quad + \quad Cl^-$$

1 mol	1 mol	1 mol
36.46 g	1.008 g	35.45 g
$6.022 * 10^{23}$	$6.022 * 10^{23}$	$6.022 * 10^{23}$

As 1 mol of HCl can produce 1 mol of H⁺, then

$$1 \text{ mol} = 1 \text{ equivalent}$$

Diprotics acids.

In diprotics acids we have:

$$H_2SO_4 \quad \rightarrow \quad 2H^+ \quad + \quad SO_4^=$$

1 mol	2 moles	1 mol
98.08 g	2 (1.008) g	96.06 g
$6.022* 10^{23}$	$2 (6.022* 10^{23})$	$6.022* 10^{23}$

1 mol can produce 2 moles of H⁺, then

$$1 \text{ mol} = 2 \text{ equivalents}$$

Triprotics acids.

For triprotics acids we have:

$$H_3PO_4 \quad \rightarrow \quad 3H^+ \quad + \quad PO_4^=$$

1 mol	3 moles	1 mol
98 g	3 (1.008) g	95 g
$6.022* 10^{23}$	$3 (6.022* 10^{23})$	$6.022* 10^{23}$

1 mol can produce 3 moles of H⁺, then

$$1 \text{ mol} = 2 \text{ equivalents}$$

Summarizing:

$$EW = \frac{MM}{N^\circ H^+}$$

B.- EQUIVALENTE WEIGHT OF A BASE:

In the case of a base with 1 hydroxyl group:

$$NaOH \quad \rightarrow \quad Na^+ \quad + \quad OH^-$$

1 mol	1 mol	1 mol
40 g	23g	17 g
$6.022 * 10^{23}$	$6.022 * 10^{23}$	$6.022 * 10^{23}$

As 1 mol of NaOH can produce 1 mol of OH⁻, then

1 mol = 1 equivalent

In the case of a base with 2 hydroxyl groups:

$$Ca(OH)_2 \quad \rightarrow \quad Ca^{2+} \quad + \quad 2OH^-$$

1 mol	1 mol	2 moles
74 g	40g	2 (17) g
$6.022 * 10^{23}$	$6.022 * 10^{23}$	$2(6.022 * 10^{23})$

As 1 mol of $Ca(OH)_2$ can produce 2 moles of OH^-, then

$$1 \text{ mol} = 2 \text{ equivalents}$$

In the case of a base with 3 hydroxyl groups:

$$Al(OH)_3 \;\; \rightarrow \;\; Al^{+3} \;\; + \;\; 3OH^-$$

1 mol	1 mol	3 moles
78 g	27 g	3 (17) g
$6.022* 10^{23}$	$6.022* 10^{23}$	$3(6.022* 10^{23})$

As 1 mol of $Al(OH)_3$ can produce 3 moles of OH^-, then

$$1 \text{ mol} = 3 \text{ equivalents}$$

Summarizing:

$$EW = \frac{MM}{N^\circ OH^-}$$

C.- EQUIVALENT WEIGHT IN REDOX REACTIONS:

In redox reactions, calculate the equivalent weight of an oxidizing substance or a reducing substance, assessing the electrons transferred in half reaction, leaving the expression as:

$$EW = \frac{MM}{N° e^-}$$

Solved problems

a) What is the normality of a solution containing 4.42 g of HNO$_3$ in 500 mL of solution?.

Resolution:

First, it is necessary to know the equivalent weight of HNO$_3$:

$$EW = \frac{MM}{1\, H^+}$$

Knowing that:

$$N° \, equivalents = \frac{g}{EW} = \frac{4.42g}{63\, g/eq} = 0.0702eq$$

To calculate Normality, we use equation:

$$N = \frac{N° equivalents}{V} = \frac{0.0702\ eq}{0.5\ L} = 0.1404\ eq/L$$

b) What is the normality of a solution of 8.57 g of barium hydroxide in 200 mL of solution? Also calculate Molarity.

Resolution:

$$\text{Equivalent Weight of Ba(OH)}_2 = \frac{MM}{N^\circ\,OH^-} = \frac{171.4g\,/\,mol}{2} = 85.7g\,/\,eq$$

Equivalents number is calculated by:

$$N^\circ eq = \frac{g}{EW} = \frac{8.57\,g}{85.7\,g\,/\,eq} = 0.1000\,eq$$

Having calculated the number of matches, proceed to calculate the Normality:

$$N = \frac{N^\circ\,equivalents}{V} = \frac{0.1000\,eq}{0.200\,L} = 0.0500\,eq\,/\,L$$

Normality of solution is 0.0500 eq/L

To calculate the Molarity, we have the following expression:

$N = n \times M$ Where n is the number of OH⁻ (in the case of a base) or the Nº of H⁺(in the case of an acid).

In this example, n = 2, because the base can give 2 OH⁻, Molarity therefore is:

$$M = \frac{0.0500\,eq\,/\,L}{2} = 0.025\,moles\,/\,L$$

1. Calculate the mass of a solution of $Ca_3(PO_4)_2$ to 15 % containing 60 grams of $Ca_3(PO_4)_2$.

ANSWER. 400 grams of solution

2. Calculate the mass of $Sn(SO_4)_2$ in 200 mL of a solution of tin sulfate IV , 10% whose density is 1.10 g / mL at a temperature of 25 ° C.

ANSWER. 22 grams

3. Determine Molarity of a solution of H3PO4 prepared by dissolving 49 grams of H_3PO_4 in 2 liters of solution.

ANSWER. 0,25 M

4. If you dissolve 28 grams of $Pt(OH)_4$ in 90 grams of water, calculate the molality of the resulting solution.

ANSWER.   1,1825 molal

5. What is the normality of a solution of 19.5 g of H_3PO_4 in 200 mL of solution? Also calculate Molarity.

ANSWER. N= 2,9847 eq/L, M= 0,9949 moles/L.

6. How many grams of water should be used to dissolve 35 grams of $Al_2(SO_4)_3$, if you wish to prepare a 2.5 molal solution?

ANSWER. 40,94 grams of water

7. Determine the Molarity and the normality of a solution of $Fe(OH)_3$ prepared by dissolving 35 grams of ferric hydroxide in 1.5 liters of solution.

ANSWER. M= 0,2183 moles/L , N= 0,6548 eq/L

8. A concentrated aqueous solution of ammonia is 32% NH_3 by weight and density 0.95 g / mL. What is the Molarity of this solution?

ANSWER. M= 17,8824 moles/L

9. How many grams of $Sn(SO_4)_2$ is required to prepare 130 mL of tin sulfate dissolution IV 0.3 Molar?.

ANSWER. 12,1177 grams

TITRATION

The ignorant says, the wise question and reflects.

Aristóteles.

*T*itration means the process by which a known concentration dissolution added to another dissolution whose concentration is unknown, until the chemical reaction between the solutes is complete.

The dissolution whose concentration is known, is called standard dissolution, and is added slowly to a known volume of the dissolution whose concentration is unknown.

During the titration, an indicator commonly is used; these are organic dyes that change color by changing the pH of the solution, which occurs after reaching the equivalence point.

Equivalence point: in the equivalence point, the number of equivalents of the base is equal to the number of equivalents of acid. It is the point at which they are carried at the same time equivalent stoichiometric amounts of substance.

Solved problems

---

a) What volume of 0.500 M NaCl is needed to completely react with 0.200 moles of Pb(NO₃)₂?.

Resolution:

To solve the problem, the first thing to do is know the chemical reaction occurring between the two reagents and stoichiometric coefficients.

$$2NaCl + Pb(NO_3)_2 \rightarrow 2NaNO_3 + PbCl_2$$

On the stoichiometry of the reaction, 2 moles of NaCl react with 1 mol Pb(NO₃)₂

To calculate the moles of NaCl that we use:

$$N°\ moles\ NaCl = 0.200\ moles\ Pb(NO_3)_2 \times \frac{2\ moles\ of\ NaCl}{1\ mol\ Pb(NO_3)_2} = 0.400\ moles\ NaCl$$

To calculate the volume of 0.5000 M NaCl solution for those moles used the expression:

$$M = \frac{N°\ moles}{V} \Rightarrow V = \frac{N°\ moles}{M} = \frac{0.400\ moles}{0.500\ moles/L} = 0.800\ Litros$$

b) Determine the normal concentration of sulfuric acid dissolution, if 25 mL of acid were titrated with 17.5 mL of a solution of NaOH 0.5 N.

Resolution:

At the point of equivalence:

N° equivalents of acid = N° equivalents of base.

As $V \times N = N°\ equivalents \Rightarrow Va \times Na = Vb \times Nb$

The data we have are:

$Va = 25\ mL$
$Vb = 17.5\ mL$
$Nb = 0.5\ eq/L$

Then: $Na = \dfrac{Vb \times Nb}{Va} = \dfrac{17.5\ mL \times 0.5\ eq/L}{25\ mL} = 0.35\ eq/L$

Normality of sulfuric acid dissolution = 0.35 eq/L

c) The amount of chloride ion in water can be determined by titration of the sample with silver nitrate using chromate ion as an indicator (when it reaches the end point of the titration to form silver chromate is red) What mass of chloride ion is present in an aqueous dissolution if required 25.0 mL of 0.150 M silver nitrate to react with all the chlorine that exists in the sample?.

Resolution:

The reaction that takes place is:

$$AgNO_{3(ac)} \quad + \quad Cl^-_{(ac)} \quad \rightarrow AgCl_{(s)} + NO^-_{3\,(ac)}$$

To determine the number of moles of$AgNO_3$ reacted:

$$N° \, moles \; AgNO_3 = V \times M = 25.0 \; mL \times \frac{1\,L}{1000\;mL} \times 0.150\;M = 0.00375\;moles$$

By stoichiometry, 1 mol of silver nitrate reacts with 1 mol of chloride.

$$N° \, moles \; Cl^- = 0.00375 \; moles \; AgNO_3 \times \frac{1\;mol\;Cl^-}{1\;mol\;AgNO_3} = 0.00375\;moles\;Cl^-$$

With the moles of Cl⁻, and knowing that MW Cl = 35.5 g/mol, we can calculate the grams of Cl⁻.

$$N° \, moles = \frac{g}{MM} \Rightarrow g = N° \, moles \times MM = 0.00375\;moles \times 35.5\;g\,/\,mol = 0.1331\,g$$

As is 0.1331 grams of chloride ion in the sample.

d) A sample of 2.0000 grams of oxalic acid unclean, required 35.5 mL of a solution of KOH 0.600 N for complete neutralization. The sample showed no acid impurities. Calculate grams of pure oxalic acid present in the sample.

Resolution:

The reaction that takes place is:

$$2KOH \;+\; (COOH)_2 \;\rightarrow\; K_2(COO)_2 \;+\; 2H_2O$$

We can calculate the number of equivalents of KOH used in this reaction:

$$N° eq_{base} = V \times N = 35.5 \; mL \times \frac{1 \, L}{1000 \, mL} \times 0.600 \; eq/L = 0.0213 \; eq$$

As $N° eq_{ácido} = N° eq_{base}$ then N° eq acid = 0.0213 eq

However:

$$N° eq = \frac{g}{EW} \quad and \quad EW = \frac{MM}{N° H^+}$$

For oxalic acid, the N ° of H + is equal to 2, and MW = 90 g / mol:

$$EW = \frac{90 \; g/mol}{2 \; eq/mol} = 45 \; g/eq \quad y \quad g = N° eq \times EW = 0.0213 \; eq \times 45 \; g/eq = 0.9585 \; grams$$

The grams of oxalic acid present in 0.2000 grams of the impure sample = 0.9585 grams.

Problems answered

1. What volume of 0.500 M H_3PO_4 is needed to completely react with 0.200 moles of $Ca(OH)_2$?.

ANSWER. V= 266,67 mL

2. A sample of 0.35 grams of NaOH unclean, took 40.0 mL of 0.200 N H_2SO_4 solution for complete neutralization. The sample showed no acid impurities.
a) Calculate the grams of pure NaOH present in the sample.
b) Calculate the percentage of purity in the sample

ANSWER. a) 0,32 grams pure, b) 91,43 % de purity.

3. Calculate the mass of barium hydroxide to be weighed to consume 15 mL of an aqueous dissolution acetic acid concentration of 0.5 M

ANSWER. 0,6422 grams

4. Titrate 25 mL of a solution of NaOH with 0.5000 M HCl and it took 35 mL of acid solution to reach the equivalence point. What is the concentration of the dissolution valued

ANSWER. N= 0,7000 eq/L

5. Titrate 30 mL of acetic acid with 25 mL of a potassium hydroxide solution 0.3500 N. Calculate the normality of the acid solution.

ANSWER. N= 0,2917 eq/L

PREPARATION OF SOLUTIONS

A scientist must be free to raise any question, to doubt any assertion, to correct mistakes.

Julius Robert Oppenheimer.

In the preparation of solutions is necessary to consider the concentration and volume of the solution to prepare.

Based on the dissolution of a solid or a liquid in a solvent in which it is miscible, we can prepare solutions of required concentrations.

If you have a concentrated solution, and wants to prepare a diluted solution, you must take into account the final concentration of the solution to be obtained, then calculates the number of moles required for the preparation of this solution and these moles are taken from the more concentrated solution.

The final moles of solute in the solution are given by:

$$N° \, final \, moles = V_f \times M_f$$

And this number of final moles must to be equal to the initial moles used to prepare this solution.

$$N° \, initial \, moles = V_i \times M_i$$

As

$$N° \, initial \, moles = N° \, final \, moles,$$

Then

$$\boxed{V_i \times M_i = V_f \times M_f}$$

a) An experiment requires 300 mL of a solution of HNO_3 1.5 m. If there is only one solution of 5.0 M HNO_3, indicate how much volume of this solution is required to prepare the diluted solution.

Resolution:

Using equation $V_i \times M_i = V_f \times M_f$ we can calculate the volume of concentrate solution need to prepare 300 mL of dilute solution:

$$V_i = \frac{V_f \times M_f}{M_i} = \frac{300 \ mL \times 1.5 \ M}{5.0 \ M} = 90 \ mL$$

It is required 90 mL of concentrate solution to prepare 300 mL of dilute solution 1,5 M.

b) If you have a solution of 6.0 N H_2SO_4 and seeks to prepare 600 mL of a 1.0 M solution in sulfuric acid, indicate the amount to be taken of the concentrated solution to prepare the diluted solution.

Resolution:

It is necessary to know the molar concentration of concentrated sulfuric acid solution. We have $N = n \times M$ where n is the number of H^+ donated by the acid.

We know that for sulfuric acid, n = 2, so the Molarity of the concentrated solution is:

$$M = \frac{N}{n} = \frac{6.0 \; eq/L}{2 \; eq/mol} = 3 \; mol/L$$

Now let's calculate the volume of the solution should be concentrated:

$$V_i = \frac{V_f \times M_f}{M_i} = \frac{600 \; mL \times 1.0 \; M}{3 \; M} = 200 \; mL$$

The volume of concentrated sulfuric acid solution to be taken to prepare the diluted solution is 200 mL.

c) There are two solutions of HNO_3, one is 5.0 M and the other is 2.0 M. Calculate the volume to be taken from each solution to make 1 liter of a solution 3.0 M.

Resolution:

To resolve this problem we must know the number of moles of HNO_3 present in the final solution:

$$N^\circ moles \; HNO_{3_f} = V_f \times M_f = 3 \; moles$$

Let A be the number of moles of the solution from 5.0 M and B is the number of moles of the solution 2.0 M, the total number of moles is equal to the sum of the moles from the 5.0 M solution and moles from of 2.0 M solution:

$$A + B = 3 \; moles$$

Now, we calculate the amount to be taken of each solution. To do this we establish that X is the volume taken from the 5.0 M solution and Y the volume taken from the solution 2.0 M.

$$X + Y = 1 \text{ Liter}$$

We can now express the number of moles to be taken of each solution, depending on the volume and Molarity:

For solution 5.0 M will be: $N^{\circ} moles = V \times M \Rightarrow A = X \times 5$

For solution 2.0 M will be $N^{\circ} moles = V \times M \Rightarrow B = Y \times 2$

Leaving two equations, with two unknowns:

$$5X + 2Y = 3 \quad \text{and} \quad X + Y = 1$$

Replacing:

$5(1 - Y) + 2\ Y = 3$ reordering:

$5 - 5Y + 2Y = 3$

being

$Y = 2/3 \text{ L} \quad \text{and} \quad X = 1/3 \text{ L}$

Problems answered

1.- What volume of a solution of 2.5 M NH_3 concentration must be taken to prepare 2.5 liters of a dilute solution with a concentration of 1 M.

ANSWER. V= 1 Liter.

2.- You want to prepare a 1.5 M solution of copper sulphate from the solid salt. Say how many grams should be weighed to prepare 0.5 liters of solution.

ANSWER. 119,63 grams

3.- If you have 1 liter of a 4.3 M solution of calcium carbonate and wants to make 4 liters of a dilute solution with a concentration of 1.0 M. Say the volume to be used in the concentrated solution to prepare the diluted solution.

ANSWER. 930,23 mL

4.- A student took 25 mL of 3.6 M NaOH solution and prepared 2 liters of a diluted solution. Tell the concentration of the solution.

ANSWER. 0.045 M

5.- If you have a solution A of 1 M HCl and a solution B concentration 2 M HCl, and you want to prepare solution of 1.7 M HCl from the two previous solutions (without adding water), say the volume to be taken from each solution.

ANSWER. Volume A= 0,3 L , Volume B= 0,7 L.

STOICHIOMETRY OF MIXTURES

Blessed are those who tasted failure, because they recognize their friends.

Joan Manuel Serrat.

A mixture is the aggregation of several substances or bodies that do not combine chemically with each other.

The mixtures are characterized by:

Its members do not lose their properties

The components involved in varying proportions.

The components can be separated by physical or mechanical.

Heterogeneous mixtures are generally (though there may be a homogeneous mixture).

Percent purity

The percentage is given by the expression:

$$\% \ purity = \frac{pure \ grams}{impure \ grams} \times 100$$

a) A sample of 3.00 grams of $CaCO_3$ impure had a purity of 90.5%. Calculate grams of pure $CaCO_3$ in the sample.

Resolution:

Using the above equation:

$$\% \ purity = \frac{pure \ grams}{impure \ grams} \times 100 \quad \Rightarrow \quad g \ pure \ CaCO_3 = \frac{\% \ purity \times g \ impure}{100}$$

$$g \ pure = \frac{90.5 \ \% \times 3.00 \ g}{100} = 2.715 \ grams \ pure \ CaCO_3$$

b) A sample of 3 grams of impure sodium chloride, dissolved in water and silver nitrate was added in excess. If precipitated 7.54 g of silver chloride. Determine the percentage of chlorine in the sample.

Resolution:

The balanced equation is:

$$NaCl \quad + \quad AgNO_3 \quad \rightarrow \quad AgCl \downarrow \quad + \quad NaNO_3$$

You must know the molecular weight of silver chloride.

MW AgCl=143.34 g/mol,   AW Cl=35.5 g/mol,   AW Ag= 107.88 g/mol

To know the grams of chlorine present in 7.54 g of AgCl and therefore the grams of chlorine in the sample, use the following relationship:

$$g\ Cl = 7.54\ g\ of\ AgCl \times \frac{35.5\ g\ of\ Cl}{143.34\ g\ of\ AgCl} = 1.87\ g\ of\ Cl$$

And now with the grams of chlorine, we can calculate the percentage of chlorine in the sample:

$$\%\ Cl = \frac{g\ Cl}{grams\ of\ sample} \times 100 = \frac{1.87\ g\ of\ Cl}{3\ g\ sample} \times 100 = 62.25\%$$

c) A sample of 0.40 g of a mixture containing only K_2SO_4 and Na_2SO_4 was dissolved and precipitated sulphate in the form of barium sulfate. The sample was filtered and dried and found that the weight of $BaSO_4$ was 0.61 g. Calculate the percentage of sodium sulfate and potassium sulfate in the sample.

Resolution:

The chemical reactions that take place are:

$$K_2SO_4 \ + \ Ba^{+2} \rightarrow \ BaSO_4 \downarrow \ + \ 2K^+$$

$$Na_2SO_4 \ + \ Ba^{+2} \ \rightarrow \ BaSO_4 \downarrow \ + \ 2Na^+$$

We see that all the barium sulfate precipitate from potassium sulfate and sodium sulfate, and the weight of the sample is given only for these compounds, therefore we can write the following equations:

1   Weight K_2SO_4 + Weight Na_2SO_4 = 0.50 g

2   Weight $BaSO_4$ (from Na_2SO_4) + Weight $BaSO_4$ (from $K_2SO_{4)}$= 0.61

By stoichiometry, we can say that 1 mol of Na_2SO_4 produces 1 mole of $BaSO_4$, and using the equation: $N°\ moles = \frac{g}{PM}$, you can calculate the grams of barium sulfate produced by the grams of sodium sulfate in the mixture, then the relationship is:

$$1 \times MW\ Na_2SO_4 \quad produce \quad 1x\ MW\ of\ BaSO_4$$

Similarly we can write:

$$1 \times MW\ K_2SO_4\ produce \quad 1x\ MW\ of\ BaSO_4$$

Therefore we can write equation 2 as:

$$Weight\ Na_2SO_4 \times \frac{MW\ BaSO_4}{MW\ Na_2SO_4} + Weight\ K_2SO_4 \times \frac{MW\ BaSO_4}{MW\ K_2SO_4} = 0.61$$

We know that the molecular weights are:

MW $BaSO_4$ = 233.4 g/mol, MW Na_2SO_4=142 g/mol, MW K_2SO_4=174.2 g/mol

Substituting in this equation the values of molecular weights of each substance we have:

$$\boxed{2} \quad Weight\ Na_2SO_4 \times \frac{233.4\ g/mol}{142\ g/mol} + Weight\ K_2SO_4 \times \frac{233.4\ g/mol}{174.2\ g/mol} = 0.61$$

From Equation 1 we have:

$\boxed{1}$ Weight Na_2SO_4 = 0.40 – Weight K_2SO_4

Substituting 1 in 2 is:

$$(0,40 - Weight\ K_2SO_4) \times \frac{233.4\ g/mol}{142\ g/mol} + Weight\ K_2SO_4 \times \frac{233.4\ g/mol}{174.2\ g/mol} = 0.61$$

Solving:

$$0.40 \times \frac{233.4 g / mol}{142 g / mol} - weight \ K_2SO_4 \times \frac{233.4 g / mol}{142 g / mol} + weight \ K_2SO_4 \times \frac{233.4 \ g / mol}{174.2 \ g / mol} = 0.61$$

$$0.656 - Weight \ K_2SO_4 \times 1.64 + Weight \ K_2SO_4 \times 1.34 = 0.61$$

$$(0.656 - 0.61) = Weight \ K_2SO_4 \times (1.64 - 1.34)$$

$$Weight \ K_2SO_4 = \frac{0.046}{0.3039} = 0.1513 \ grams$$

From equation 1:

Weight Na_2SO_4 = 0.40 - 0.1513= 0.2486 $grams$.

d) It was a weighed sample of 0.20 grams of a mixture containing only sodium chloride and sodium bromide, with a silver nitrate dissolution and obtained a precipitate of 0.40 grams. Calculate the percentage of sodium chloride and sodium bromide in the mix.

Resolution:

Denote by X and Y quantities of sodium chloride and sodium bromide in the mixture, then:

1   X + Y = 0.20

During precipitation, the following reactions occur:

$$NaCl \ + \ AgNO_3 \rightarrow \ AgCl \downarrow \ + \ NaNO_3$$

$$NaBr \ + \ AgNO_3 \rightarrow \ AgBr \downarrow \ + \ NaNO_3$$

The molecular weights of the compounds involved in the reaction are:

MW AgCl= 143.32 g/mol, MW NaCl=58.44 g/mol, MW Ag Br= 187.78 g/mol, MW NaBr = 102.90 g/mol.

The grams of precipitate are of NaCl and NaBr, as is the equation 2:

2 $Weight \ AgCl + Weight \ AgBr = 0.40$

To calculate the grams of AgCl from the NaCl, we have:

$$Weight \ AgCl \times \frac{1 \ MW \ AgCl}{1 \ MW \ NaCl} = X \times \frac{143.32 \ g \ / \ mol}{58.44 \ g \ / \ mol} = X \times 2.45$$

Similarly, to calculate the grams of AgBr coming from NaBr:

$$Weight\ AgBr \times \frac{1\ MW\ AgBr}{1\ MW\ NaBr} = Y \times \frac{187.78\ g/mol}{102.90\ g/mol} = Y \times 1.82$$

We can now write two equations with two unknowns:

1                 $X + Y = 0.20$

2                 $2.45\ X + 1.82\ Y = 0.40$

Substituting 1 in 2 we have:

$2.45\ (0.20-Y) + 1.82\ Y = 0.40$ solving:

$0.49 - 2.45\ Y + 1.82\ Y = 0.40$ and $(0.49-0.40) = (2.45-1.82)\ Y$

$Y = 0.14$ grams $\Rightarrow$ There are 0.14 grams of NaBr

$X = 0.06$ grams $\Rightarrow$ There are 0.06 grams of NaCl

To calculate the percentages:

$$\%\ NaBr = \frac{grams\ NaBr}{gramo\ sample} \times 100 = \frac{0.14\ g}{0.20\ g} \times 100 = 70\ \%\ NaBr$$

$$\%\ NaCl = \frac{grams\ NaCl}{grams\ sample} \times 100 = \frac{0.06\ g}{0.20\ g} \times 100 = 30\ \%\ NaCl$$

e) A mixture of 0.40 grams, containing KCl and KI, was treated with a solution containing sulfate ions and after drying was obtained 0.25 grams of potassium sulphate. Determine the grams of KCl and KI in the mixture.

Resolution:

To resolve this problem, again we use a system of equations with two unknowns. Let X and Y grams of KCl the gram of KI in the mixture, then equation 1 is:

1   X + Y = 0.40

The chemical reactions that take place are:

$$2KCl + SO_4^= \rightarrow K_2SO_4 + 2Cl^-$$

$$2KI + SO_4^= \rightarrow K_2SO_4 + 2I^-$$

We now have our second equation:

2   Weight K_2SO_4 (from KCl) + Weight K_2SO_4 (from KI) = 0.25

To determine the grams of K_2SO_4 from the KCl must be by stoichiometry, 2 moles of KCl produces 1 mol of K_2SO_4.

Then: Weight K_2SO_4 (from KCl) = $\dfrac{X \times 1\, MM K_2SO_4}{2\, MM KCl}$

Analogously,   Weight K_2SO_4 (from KI) = $\dfrac{Y \times MM\ K_2SO_4}{2\, MM\ KI}$

Substituting in equation 2 we have:

$$2 \quad X \times \frac{1 \; MW \; K_2SO_4}{2 \; MW \; KCl} \quad + \quad Y \times \frac{1 \; MW \; K_2SO_4}{2 \; MW \; KI} = 0.25$$

The molecular weights of the substances involved in the reactions are:

MW KI= 166.01 g/mol,   MW KCl = 74.56 g/mol,   MW K_2SO_4= 174.27 g/mol.

Substituting in equation 2 we have:

$$2 \quad X \times \frac{174.27 g / mol}{2 \times 74.56 g / mol} \quad + \quad Y \times \frac{174.27 g / mol}{2 \times 166.01 g / mol} = 0.25$$

Solving: X x 1.17 + Y x 0.52 = 0.25

Substituting equation 1 into equation 2 :

(0.40-Y) x1.17   +   Y x 0.52 = 0.25

Solving: (0.47 - 0.25) = (1.17 - 0.52) Y

Y= 0.22/0.65 = 0.34 $\Rightarrow$ grams of KCl = 0.34 grams

X= 0.06 $\Rightarrow$ grams of KI = 0.06 grams

f) The CO_2 content of a mixture of calcium carbonate and strontium carbonate is 35%. Determine the percentage of each component in the mixture.

Resolution:

Again, we resorted to the use of two equations with two unknowns, and we establish that X is the percentage of $CaCO_3$ in the mixture and Y is the percentage of $SrCO_3$, leaving the equation 1 as:

1   X + Y = 100 %

And equation 2 as:

2   a + b = 35 %

It is known that in determining the content of CO_2 in $CaCO_3$ is used to:

$$CaCO_3 \rightarrow CO_2$$

And similarly for $SrCO_3$

$$SrCO_3 \rightarrow CO_2$$

The molecular weights of the compounds are:

MW CO_2= 44 g/mol,   MW $CaCO_3$= 100 g/mol,   MW $SrCO_3$ = 147.62 g/mol

So to calculate the CO_2 content in the components of the mixture we have:

$$\% \ CO_2 \ in \ CaCO_3 = 100 \times \frac{1 \ MW \ CO_2}{1 \ MW \ CaCO_3} = 100 \times \frac{44 \ g/mol}{100 \ g/mol} = 44 \ \% \ CO_2 \ in \ CaCO_3$$

% CO_2 in $SrCO_3$=

$$100 \times \frac{1 \ MW \ CO_2}{1 \ MW \ SrCO_3} = 100 \times \frac{44 g/mol}{147.62 \ g/mol} = 29.81 \ \% \ CO_2 \ in \ SrCO_3$$

Now, to determine the CO_2 content in each component of the mixture:

In $CaCO_3$: $\quad g \ CO_2 \ in \ CaCO_3 = a = X \times \dfrac{44}{100}$

In $SrCO_3$: $\quad g \ CO_2 \ in \ SrCO_3 = b = Y \times \dfrac{29.81}{100}$

From Equation 2, we have: $\qquad a \ + \ b \ = \ 35 \ \%$

Substituting in this expression: $\qquad X \times \dfrac{44}{100} \ + \ Y \times \dfrac{29.81}{100} = 35 \ \%$

Is the equation 2 as: $\quad 0.44 \ X \ + \ 0.2981 \ Y \ = 35 \ \%$

Substituting equation 1 in 2:

$0.44 \ (100 \ \% - Y) \ + \ 0.2981 \ Y \ = \ 35 \ \%$

Solving: $\qquad (44 \ \% - 35 \ \%) \ = \ Y \ (0.44 - 0.2981)$

$Y = 63.42 \ \% \quad y \quad X = 36.58 \ \%$

In the sample there are 63.42 % of $CaCO_3$ and 36.58 % of $SrCO_3$.

1.- How many grams of pure NaOH are present in a sample of 5.0 grams of NaOH at 95% purity?

ANSWER. 4,75 grams of NaOH pure.

2.- A sample of potassium chlorate and potassium chloride weighs 50 grams. After prolonged heating of the sample, is obtained 4.8 grams of Oxygen. Calculate the grams of each compound in the original mixture.

ANSWER. Potassium chlorate = 12,5 grams;   potassium chloride = 37,5 grams

3.- A mixture of potassium iodide (MW = 213.9) and potassium chloride (MW = 74.6) that weighs 10 grams, is dissolved in water to make a dissolution of 1.0 L. Calculate the amount of each of the initial compounds when 25 mL of the previous dissolution have required 26.2 mL of $AgNO_3$ (MW = 169.9) 0.1 M for complete precipitation

ANSWER. Potassium iodide = 2,182 grams; potassium chloride =7,818 grams

4.- A sample of 2 g composed of SnO and SnO_2, descomposes with heat. Tin formed weighs 1.66 grams. Calculate the percentage of SnO in the mix. Atomic weight: Sn = 118.7, O = 16

ANSWER. SnO_2 = 45.66 % ;   SnO = 54,34 %.

REDOX REACTIONS

Science is made up of mistakes, which in turn, is steps towards the truth.

Julio Verne.

Oxide reduction reactions are those, in which the atoms undergo change in oxidation number, therefore there is a transfer of electrons.

Balancing redox reactions

In balancing the reactions of oxide reduction (redox) is necessary to know the following terms:

Oxidation: is an increase in oxidation number, which corresponds to the loss of electrons.

Reduction: is a decrease in oxidation number, which corresponds to the gain of electrons.

Oxidation State: is the net charge of atom, e.g. Cl⁻, has an oxidation state of -1, Co^{+2} has a +2 oxidation state.

An example of an *oxidation reaction* is:

$$2\overset{+2\ -2}{C\,O}_{(g)} + \overset{0}{O}_{2(g)} \rightarrow 2\overset{+4\ -2}{C\,O}_{2(g)}$$

Carbon In this reaction goes from an oxidation state +2 to +4 oxidation state, therefore oxidized. As an example of a *reduction reaction* we have:

$$\overset{+2}{Cu} + Zn \leftrightarrow \overset{0}{Cu} + \overset{+2}{Zn}$$

In this case, the Cu pass from one oxidation state to +2 oxidation state 0, is therefore reduced.

An example of an *oxide reduction reaction* can be viewed at:

$$2\,\overset{0}{Fe}_{(s)} + 3\,\overset{0}{Cl}_{2(g)} \rightarrow 2\,\overset{+3}{Fe}\overset{-1}{Cl}_{3(s)}$$

In this reaction iron goes from oxidation state 0 to +3, it is oxidized, and, the chlorine pass from one oxidation state 0 to an oxidation state of -1, therefore is reduced.

In the reactions of oxide reduction, the species that is oxidized (lose electrons) is called the *Reducing agent,* and the species is reduced (gains electrons) is called the *Oxidizing agent.*

Rules for assigning oxidation states in molecules

The oxidation state of the pure elements in any allotrope is zero (0).

The oxidation state of Hydrogen is +1 is all made up, except for the hydrides, in which case it is -1

The oxidation state of Oxygen is -2 in all compounds except peroxides (-1) (H_2O_2) or super-oxides (- ½) (KO_2).

Fluorine has oxidation state -1 for all compounds. The halogens have oxidation state -1 exists as halide ions. When combined with Oxygen has positive oxidation state.

In a neutral molecule, the sum of oxidation numbers of all atoms must be zero.

p. e.g.. $\overset{+1\ +7\ -2}{K\ Mn\ O_4}$ $(+1) + (+7) + (-8) = 0$

$\overset{+1\quad +6\quad -2}{K_2\ Cr_2\ O_7}$ $2(+1) + 2\ (+6) + 7(-2) = 0$

Steps for balancing oxide reduction reactions:

Assign oxidation states to all the species involved.

Identify species that are oxidized and reduced

Write semi-independent reactions for oxidation and reduction

Balancing these semi-reactions to the nuclei (atoms) and the electric charge.

Combining balanced half reactions to give rise to the net overall reaction Oxidation-Reduction.

1.- Balance by the method of electron ion in acid medium equation.

$$MnO_4^- + C_2O_4^= \longrightarrow Mn^{+2} + CO_2$$

Resolution:

a) As mentioned earlier, as a first step we assign the oxidation states of all species involved:

$$\begin{array}{cc} +7 & -2 \end{array}$$
$$MnO_4^- : MnO_4^- \quad (+7) + (-8) = -1$$

$$\begin{array}{cc} +3 & -2 \end{array}$$
$$C_2O_4^{-2} : C_2O_4^{-2} \quad (+6) + (-8) = -2$$

$$Mn^{+2}$$

$$\begin{array}{cc} +4 & -2 \end{array}$$
$$CO_2 : CO_2 \quad (+4) + (-2) = 0$$

b) The second step is to identify species that are oxidized and reduced, in this case:

Manganese valence changes from +7 to +2 valence therefore gains electrons and is reduced. One element that reduces is the *oxidizing agent*.

Carbon pass from the valence +3 to +4 valence therefore loses electrons and is oxidized. The species that is oxidized is the *reducing agent*.

c) Now that we have identified the species that are oxidized and reduced, we write the half reactions:

$$\overset{+7\ -2}{MnO_4^-} \quad \longrightarrow \quad Mn^{+2}$$

$$\overset{+3\ -2}{C_2O_4^{-2}} \quad \longrightarrow \quad \overset{+4\ -2}{CO_2}$$

d) The fourth step is to balance the half reactions both in their nuclei and in their electrical charge

$$\overset{+7\ -2}{MnO_4^-} \quad \longrightarrow \quad Mn^{+2}$$

As the side of the reactive oxygen exists, it is necessary to place the product side oxygen, and this is put into water, because as we're balancing in acidic media, hydrogen must be placed on the side of the reactants and water on the side of Products:

$$\overset{+7\ -2}{MnO_4^-} \quad \longrightarrow \quad Mn^{+2} + H_2O$$

Now, we put water on the side of products, and we must balance the masses by placing hydrogen (as H^+) in the reagents,

$$\overset{+7\ -2}{MnO_4^-} + H^+ \quad \longrightarrow \quad Mn^{+2} + H_2O$$

Once we have placed the masses, we need to balance.

As we have 4 oxygen in the reactant side must put 4 oxygen on the side of the products:

$$+7 \quad -2$$
$$MnO_4^- \;+\; H^+ \quad \longrightarrow \quad Mn^{+2} \;+\; 4\,H_2O$$

Now we have 8 hydrogen on the product side, we must place 8 hydrogen on the side of reagents

$$+7 \quad -2$$
$$MnO_4^- \;+\; 8\,H^+ \quad \longrightarrow \quad Mn^{+2} \;+\; 4\,H_2O$$

Once we have placed the masses, we must be sure they are balanced, so we review what we have in the side of the reagents and this must be equal to what we have the product side:

REAGENTS:	PRODUCTS:
1 Manganese	1 Manganese
4 Oxygen	4 Oxygen
8 Hydrogen.	8 Hydrogen.

Now that we have balanced the masses we must balance the charges:

Side of the reagents we have: (-1) + (8 +) = +7

The product side we have: +2 + (0) = +2

Therefore we must add 5 electrons (each one has a negative charge) of the reactant side so that both sides have the same charge:

$$\overset{+7 \quad -2}{MnO_4^-} + 8\ H^+ + 5\ e^- \longrightarrow Mn^{+2} + 4\ H_2O$$

Side of the reagents we have: $(-1) + (8+) + (5-) = +2$

The product side we have: $+2 + (0) = +2$

Once we have a semi balanced reaction, we must balance the other half reaction:

$$\overset{+3\ -2}{C_2O_4^{-2}} \longrightarrow \overset{+4\ -2}{CO_2}$$

On the side of the reagents have 2 Carbons, while the product side we have 1, so we must balance:

$$\overset{+3\ -2}{C_2O_4^{-2}} \longrightarrow 2\ \overset{+4\ -2}{CO_2}$$

By balancing the Carbons we must verify that the masses are balanced:

REAGENTS:	PRODUCTS:
2 Carbon	2 Carbon
4 Oxygen	4 Oxygen

Since the masses are balanced, we must now balance the load:

$$\overset{+3\ -2}{C_2O_4^{-2}} \longrightarrow \overset{+4\ -2}{2\ CO_2}$$

On the side of the reagents are 2 negative charges, on the side of products it's no charge. Balance must be added to the product side the electrons transferred during the reaction of oxide reduction. As discussed above, each Carbon went from having valence +3 to +4 valence, as we have two Carbons in the molecule, then the number of electrons transferred is 2:

$$\overset{+3\ -2}{C_2O_4^{-2}} \longrightarrow \overset{+4\ -2}{2\ CO_2} + 2\ e^-$$

Now we have the balanced half reaction.

e) As a final step we combine the two half reactions:

$$\overset{+7\ -2}{MnO_4^-} + 8\ H^+ + 5\ e^- \longrightarrow Mn^{+2} + 4\ H_2O$$

$$\overset{+3\ -2}{C_2O_4^{-2}} \longrightarrow \overset{+4\ -2}{2\ CO_2} + 2\ e^-$$

We see that the number of electrons transferred in the first half reaction of 5, while the second is 2. As the final result must be the same number in both, then we multiply top half reaction by 2 and the lower 5:

$$(\ \overset{+7\ -2}{MnO_4^-} + 8\ H^+ + 5\ e^- \longrightarrow Mn^{+2} + 4\ H_2O\ )\ x$$

2

$$(\ \overset{+3\ -2}{C_2O_4^{-2}} \longrightarrow \overset{+4\ -2}{2\ CO_2} + 2\ e^-)\ x\ 5$$

Adding the two half-reactions have

$$2MnO_4^- + 16 H^+ + 5 C_2O_4^{-2} + \cancel{10} e^- \longrightarrow 2 Mn^{+2} + 8 H_2O + 10CO_2 + \cancel{10} e^-$$

By removing electrons on both sides of the reaction is the final equation balanced:

$$2MnO_4^- + 16 H^+ + 5 C_2O_4^{-2} \longrightarrow 2 Mn^{+2} + 8 H_2O + 10CO_2$$

2.- Balance the following equation in acidic media:

$$Fe^{+2} + NO_3^- \longrightarrow Fe^{+3} + NO$$

Resolution:

a) The first step to assign oxidation states to all the species involved:

$$\overset{+5\ -2}{NO_3^-} :\ NO_3^- : (+5) + (-6) = -1 \quad Fe^{+2} \text{ has valence } +2$$

$$\overset{+2\ -2}{NO} :\ NO : (+2) + (-2) = 0$$

b) The second step is to identify species that are oxidized and reduced, in this case:

Nitrogen passes from Valencia +5 to Valencia +2 thus gains electrons and is reduced. One element that is reduced is the *oxidizing species*

Iron passes from Valencia +2 to valence +3, therefore loses electrons and is oxidized. The species that is oxidized is the *reducing agent*.

c) After identifying the species that are oxidized and reduced, we write the half reactions:

$$Fe^{+2} \longrightarrow Fe^{+3}$$

$$NO_3^- \longrightarrow NO$$

d) The fourth step is to balance the half reactions both in their nuclei and in their electrical charges:

$$Fe^{+2} \longrightarrow Fe^{+3}$$

In this half reaction, we have the same amount of iron on both sides of the equation but the charges are unbalanced. As iron became iron +2 to iron +3, this happened because it lost an electron, so the electron is added to the product side.

$$Fe^{+2} \longrightarrow Fe^{+3} + e^-$$

Moreover, in the reaction:

$$NO_3^- \longrightarrow NO$$

Nitrogen goes from +5 to +2, so winning 3 electrons. These three electrons are added to the reactant side.

$$NO_3^- + 3e^- \longrightarrow NO$$

Nitrogen review now and see that there is equal on both the amount of reagents as in the product. However, in the oxygen, we see that there are three on the side of the reactants but only 1 in the product.

Since the reaction is acidic, water then placed on the side of the products:

$$NO_3^- + 3e^- \longrightarrow NO + 3 H_2O$$

But now it is necessary to put hydrogen on the side of the reagents:

$$NO_3^- + 6H^+ + 3e^- \longrightarrow NO + 3 H_2O$$

After placing the charges and the masses, inspect both sides of the half reaction the same amount of each of the elements involved and the charges are the same:

REAGENTS:	PRODUCTS:
1 Nitrogen	1 Nitrogen
3 Oxygen	3 Oxygen
6 Hydrogen	6 Hydrogen

In relation to charges in the reactant side the load is zero and zero in the product side.

Now we have the semi-balanced reactions to the last step we:

e) In this last step we combine the two half reactions:

$$Fe^{+2} \longrightarrow Fe^{+3} + e^-$$

$$NO_3^- + 6H^+ + 3e^- \longrightarrow NO + 3H_2O$$

We see that in the first half reaction was transferred to a single electron, while the second was transferred three, so multiply the first half reaction of three and the second by one:

$$(\quad Fe^{+2} \quad \longrightarrow \quad Fe^{+3} + e^-) \times 3$$

$$(\quad NO_3^- + 6H^+ + 3e^- \quad \longrightarrow \quad NO + 3H_2O) \times 1$$

And now we add the two half reactions:

$$3Fe^{+2} + NO_3^- + 6\cancel{H^+} + 3e^- \quad \longrightarrow \quad 3Fe^{+3} + NO + 3\cancel{H_2O} + 3e^-$$

Being the final reaction of oxide reduction:

$$3Fe^{+2} + NO_3^- + 6H^{+-} \quad \longrightarrow \quad 3Fe^{+3} + NO + 3H_2O$$

We conducted two exercises balancing two oxide reduction reactions in acidic media, and we saw it necessary to add H^+ and H_2O.

Now look at two examples in which there will be a redox reaction in basic media.

3.- Balance the following equation in basic medium:

Resolution:

$$MnO_4^- + Cl^- \longrightarrow Mn^{+2} + Cl_2$$

a) As mentioned earlier, the first step is to assign oxidation states to all the species involved:

$$+7 \quad -2$$
MnO_4^- : MnO_4^- $(+7) + (-8) = -1$
Cl^- : $Cl^- = -1$

Mn^{+2} : $Mn^{+2} = +2$

Cl_2 :   in this case as the chlorine is in its elemental state, its charge is $= 0$

b) The second step is to identify species that are oxidized and reduced, in this case:

Manganese pass from valence +7 to valence +2 therefore gains electrons and is reduced. One element that reduces is the *oxidizing agent*.

Chlorine passes from valence -1 to valence 0, therefore loses electrons and is oxidized. The species that is oxidized is the *reducing agent*.

c) Now that we have identified the species that are oxidized and reduced, we write the half reactions:

$$MnO_4^- \longrightarrow Mn^{+2}$$

$$Cl^- \longrightarrow Cl_2$$

d) The fourth step is to balance the half reactions both in their nuclei and in their electrical charges:

$$MnO_4^- \longrightarrow Mn^{+2}$$

As we are working in a basic medium, we will work with OH^- and H_2O:

Since manganese gained 5 electrons, those electrons placed on the side of reagents:

$$MnO_4^- + 5e^- \longrightarrow Mn^{+2}$$

Since we have six negative charges on the reactant side, we will post the OH^- on the product side to balance the half reaction and the water in the reactant side:

$$MnO_4^- + H_2O + 5e^- \longrightarrow Mn^{+2} + OH^-$$

Now we balance the charges and masses:

$$MnO_4^- + 4H_2O + 5e^- \longrightarrow Mn^{+2} + 8OH^-$$

With respect to the other half reaction

$$Cl^- \longrightarrow Cl_2$$

As every atom lost an electron, we are going to balance the half reaction as follows:

$$2Cl^- \longrightarrow Cl_2 + 2\,e^-$$

The charges are being balanced as the masses.

e) The last step is to combine the two half reactions:

$$MnO_4^- + 4\,H_2O + 5\,e^- \longrightarrow Mn^{+2} + 8\,OH^-$$

$$2Cl^- \longrightarrow Cl_2 + 2\,e^-$$

In order to have an equal number on both sides of the reaction, multiply the first half reaction by 2 and the second by 5:

$$(MnO_4^- + 4\,H_2O + 5\,e^- \longrightarrow Mn^{+2} + 8\,OH^-)\; x\;2$$

$$(2Cl^- \longrightarrow Cl_2 + 2\,e^-)\; x\;5$$

_____

Adding we have:

$$2\,MnO_4^- + 8\,H_2O + 10Cl^- + \cancel{10}\,e^- \longrightarrow 2\,Mn^{+2} + 16\,OH^- + 5\,Cl_2 + \cancel{10}\,e^-$$

And we have the final result:

$$2\,MnO_4^- + 8\,H_2O + 10Cl^- \longrightarrow 2\,Mn^{+2} + 16\,OH^- + 5\,Cl_2$$

Redox Titrations

Now that we balance oxide reduction reactions, we can determine the concentration of an unknown solution from another of known concentration using the equations balanced.

For this, we will work using the following equations:

1.- To work with Normality:

   N° equivalents of oxidizing agent = N° equivalents of reducting

$$\boxed{V \times N = N^\circ \, equivalents \implies Vo \times No = Vr \times Nr}$$

2.- To work with Molarity:

$$V \times M = N^{\circ} moles$$

a) A completely reacted dissolution of 1.5 M potassium permanganate with 200 mL of ferrous chloride dissolution 2 N according to the reaction::

$$MnO_4^- + Fe^{+2} \longrightarrow Fe^{+3} + Mn^{+2} \ (acid\ media)$$

1.- Balance by the method of ion-electron

2.- What volume of 1.5 M potassium permanganate was required for complete reaction.

Resolution:

To solve the equation we need to identify the species that change their oxidation state, in this case is the Iron and Manganese.

$$\overset{+7\ -2}{MnO_4^-} + \overset{+2}{Fe^{+2}} \longrightarrow \overset{+3}{Fe^{+3}} + \overset{+2}{Mn^{+2}}$$

Once identified we write the half reactions and we balanced:

$$Fe^{+2} \longrightarrow Fe^{+3} + e^-$$

$$MnO_4^- + 8\,H^+ + 5e^- \longrightarrow Mn^{+2} + 4\,H_2O$$

Multiply the half reactions to equal the number of electrons on both sides of the reaction and have the final equation balanced:

$$MnO_4^- + 5\,Fe^{+2} + 8\,H^+ \longrightarrow 5\,Fe^{+3} + Mn^{+2} + 4\,H_2O$$

To resolve the problem we can work with either Molarity or normality. In this exercise we will work with both.

i) **Working with Normality**:

To work with normality we must to know the Molarity of potassium permanganate:

We know that its Molarity is 1.5 moles / liter.

In the semi-reduction reaction MnO_4^- exchange the 5 electrons, thus to determine the normal use equation

$$EW = \frac{MW}{N^\circ e^-}$$

Therefore the equivalent weight is equal to the Molecular Weight / 5,

Using equation:

$$N = n \times M$$

We calculate Normality:

N= 5 x 1.5 moles/L = 7.5 eq/L

As

$$V \times N = N° \, equivalents \Rightarrow \quad Vo \times No = Vr \times Nr$$

Vo x No = N° equivalents oxidant

N° equivalents oxidant = N° equivalents reducer

N° equivalents reducting = Vr x Nr

N° equivalents reducer = 0.2 L x 2 N = 0,4 eq.

N° eq oxidant = 0,4 eq = Vo x No   and Vo= N° eq/ No = 0,4/ 7,5 eq/L

Vo = 0,0533 L

Answer: It took 0,0533 Liters of $KMnO_4$ to react completely with 200 mL of ferrous chloride dissolution 2 N

ii) **Working with Molarity**:

To work with Molarity we have to know stoichiometry of the reaction:

We know that 1 mole of $KMnO_4$ reacts with 5 moles of ferrous chloride.

We calculate the number of moles of ferrous chloride which react with potassium permanganate.

Using equation:

$$N = n \times M$$

And knowing by the balance of the half reaction, the number of electrons transferred is 1

Molarity of $FeCl_2$ is M= N/n = 2 M.

The number of moles of ferrous chloride is:

$$Vr \times Mr = 0{,}200 \text{ L} \times 2 \text{ M} = 0{,}4 \text{ moles.}$$

Using the stoichiometric ratio, then you need 0.08 moles of potassium permanganate to react with 0.4 moles of ferrous chloride.

As the number of moles of $KMnO_4$ = V × M

Then V = N° moles/M= 0,08/ 1,5 = 0, 0533 L of $KMnO_4$ 1.5 M.

b) **500 mL of a solution $Cr_2(SO_4)_3$ 2,0 N is mixed with 150 mL $KClO_3$ and reaction is:**

$$Cr_2(SO_4)_3 + KClO_3 + KOH \rightarrow K_2CrO_4 + KCl + K_2SO_4 + H_2O$$

1.- Balance by the method of electron ion.

2.- Determine Normality of $KClO_3$ solution

Resolution:

We see that this reaction occurs in basic media.

To solve the equation we need to identify the species that change their oxidation state, in this case is the Chromium and chlorine.

$$\text{2 Cr}^{+3} \longrightarrow \overset{+6\ -2}{\text{CrO}_4^{-2}} + 6\,e^-$$

$$\overset{+5\ -2}{\text{ClO}_3^-} + 6\,e^- \longrightarrow \text{Cl}^-$$

Once identified we write the half reactions and we balanced:

$$\text{2 Cr}^{+3} + 16\ \text{OH}^- \longrightarrow 8\ \text{H}_2\text{O} + 2\ \overset{+6\ -2}{\text{CrO}_4^{-2}} + 6\,e^-$$

$$\overset{+5\ -2}{\text{ClO}_3^-} + 6\,e^- + 3\ \text{H}_2\text{O} \longrightarrow \text{Cl}^- + 6\ \text{OH}^-$$

Add the half reactions to equal the number of electrons on both sides of the reaction and have the final equation balanced:

$$\text{Cr}_2(\text{SO}_4)_3 + 10\ \text{KOH} + \text{KClO}_3 \rightarrow 2\ \text{K}_2\text{CrO}_4 + 5\ \text{H}_2\text{O} + \text{KCl} + 3\ \text{K}_2\text{SO}_4$$

To resolve the problem we can work with either Molarity or normality. In this exercise we will work with teaching purposes both.

i) **Working with Normality**:

To work with Normality we use equation:

$$\boxed{V \times N = N^\circ\,equivalents \Rightarrow Vo \times No = Vr \times Nr}$$

N KClO_3 = V $\text{Cr}_2(\text{SO}_4)_3$ x N $\text{Cr}_2(\text{SO}_4)_3$ /V KClO_3

N KClO_3 = 6,6666 N

Answer: Normality of $KClO_3$ is 6,6666 eq/L.

ii) **Working with Molarity**:

To work with Molarity, we see the reaction stoichiometry:

We know that 1 mol of $Cr_2(SO_4)_3$ reacts with 1 mol of $KClO_3$.

To calculate the Molarity of both species involved in the reaction of oxide reduction we use the formula:

$$\boxed{N = n \times M}$$

In the case of chromium III sulfate, the number of electrons transferred is 6, so $M = N/n = 2/6 = 0,3333$ moles/L

The number of moles reacted chromium III sulfate is:

$$N° \text{ moles} = V \times M = 0,500 \text{ L} \times 0,3333 \text{ moles/L} = 0,1666 \text{ moles.}$$

As for the stoichiometry of reaction, 1 mol of chromium III sulfate reacts with 1 mol of $KClO_3$, then 0.1666 moles have reacted to this compound.

The volume was reacted: 0.150 L, so the Molarity is:

$V \times M = N° \text{ moles} \quad M = N° \text{ moles}/V = 0,1666 \text{ moles}/0,150 \text{ L} = 1,1111 \text{ moles/L}$

To calculate Normality we apply:

$$\boxed{N = n \times M}$$

We have seen that the average reaction is transferred, 6 electrons, so the normality of $KClO_3$ is:

$$N = 6 \text{ eq / mol} \times 1.1111 \text{ mol / L} = 6.6666 \text{ eq / L}$$

This value is equal to that obtained working normally.

1.- Balance by the method of electron ion the following reactions in acidic media:

a) $Zn + NO_3^- + H^+ \longrightarrow Zn^{+2} + NH_4^+ + H_2O$

Answer: $4\,Zn + NO_3^- + 10\,H^+ \longrightarrow 4\,Zn^{+2} + NH_4^+ + 3\,H_2O$

b) $Cu + HNO_3 \dashrightarrow NO_2 + H_2O + Cu(NO_3)_2$

Answer: $Cu + 4\,HNO_3 \longrightarrow 2\,NO_2 + 2\,H_2O + Cu(NO_3)_2$

c) $KMnO_4 + H_2SO_4 + KI \longrightarrow MnSO_4 + I_2 + K_2SO_4 + H_2O$

Answer:

$2\,KMnO_4 + 8\,H_2SO_4 + 10\,KI \longrightarrow 2\,MnSO_4 + 5\,I_2 + 6\,K_2SO_4 + 8\,H_2O$

2.- Balance by the method of electron ion the following reactions in basic media:

a) $ClO_3^- + I^- \longrightarrow Cl^- + I_2$

Answer: $ClO_3^- + 6\,I^- + 3\,H_2O \longrightarrow Cl^- + I_2 + 6\,OH^-$

b) $CrO_4^{-2} + Fe(OH)_2 \longrightarrow CrO_2^{-2} + Fe(OH)_3$

Answer: $CrO_4^{-2} + 4\,Fe(OH)_2 + 2\,H_2O \longrightarrow CrO_2^{-2} + 4\,Fe(OH)_3$

c) $\qquad$ $Fe(OH)_2 + H_2O_2 \longrightarrow Fe(OH)_3 + H_2O$

Answer: $\qquad$ $2\,Fe(OH)_2 + H_2O_2 \dashrightarrow 2\,Fe(OH)_3$

3.- **To determine the concentration of a solution of Fe $^{+2}$, a volume of 50 mL containing Fe $^{+2}$ react with 25 mL of a solution of 0.03 M $K_2Cr_2O_7$ in acid medium.**

a) **Balance the reaction by the method of electron-ion.**

b) **Calculate the molar concentration of Fe $^{+2}$.**

Answer:

a) $\qquad$ $Cr_2O_7^{-2} + 6\,FE^{+2} + 14\,H^+ \longrightarrow Cr^{+3} + 6\,FE^{+3} + 7\,H_2O$

b) $\qquad$ $[\,FE^{+2}\,] = 0{,}09\ M$

4.- **Mix one volume of 35 mL of iodine solution with 20 mL. of potassium nitrate 3N and with 15 mL of sulfuric acid 4.0 N, to react in accordance with the following:**

$$KNO_3 + I_2 + H_2SO_4 \rightarrow KIO_3 + NO$$

a) **Balance the reaction by the method of electron-ion.**

b) **Calculate the Normality of iodine.**

Answer:

a) $\quad 10\,KNO_3 + 3\,I_2 + 2\,H_2SO_4 \rightarrow 6\,KIO_3 + 10\,NO + 2\,H_2O + 2\,K_2SO_4$

b) $\ N = 1{,}7143\ eq/L$

5.- **10 grams of solid PbS react with 100 mL of a solution of Hydrogen peroxide according to reaction:**

$$PbS \text{ (s) } + H_2O_2 \rightarrow PbSO_4 + H_2O$$

a) **Balance the reaction by the method of electron-ion.**

b) **Calculate the Normality of Hydrogen peroxide.**

Answer:

$$2 \, PbS \; + \; 24 \; H_2O_2 \; \rightarrow \; 2 \, PbSO_4 + 24 \, H_2O + 8 \, O_2$$

N =3,3445 eq/L

Periodic Chart of Elements

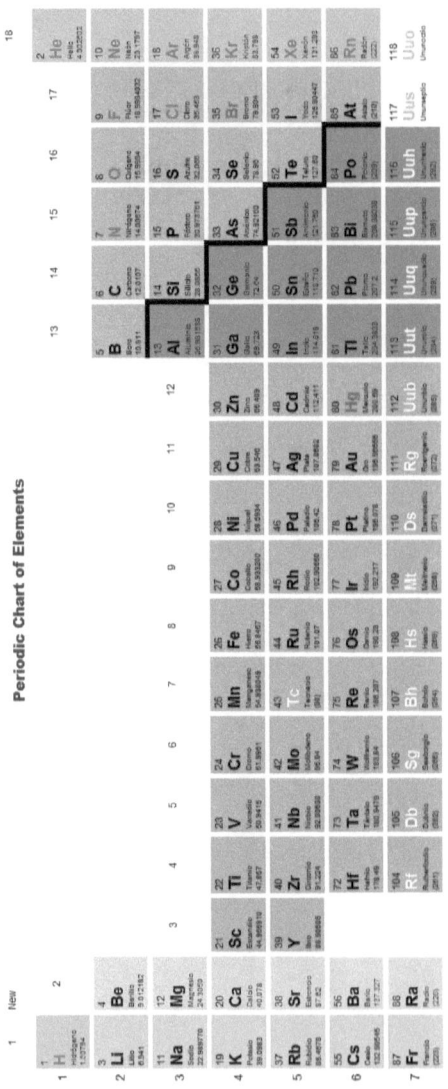

ATOMIC MASS OF THE ELEMENTS

Element	Symbol	Atomic Number	Atomic Mass
Actinium	Ac	89	(227)
Aluminum	Al	13	26,981539
Americium	Am	95	(243)
Antimonies	Sb	51	121,75
Argon	Ar	18	39,948
Arsenic	As	33	74,92159
Astatine	At	85	(210)
Sulfur	S	16	32,066
Barium	Ba	56	137,237
Beryllium	Be	4	9,012182
Berkelium	Bk	97	(247)
Bismuth	Bi	83	208,98037
Boron	B	5	10,811
Bromine	Br	35	79,904
Cadmium	Cd	48	112,411
Calcium	Ca	20	40,078
Californium	Cf	98	(251)
Carbon	C	6	12,011
Cerium	Ce	58	140,115
Cesium	Cs	55	132,90543
Chlorine	Cl	17	35,4527
Cobalt	Co	27	58,93320
Copper	Cu	29	63,546
Chromo	Cr	24	51,9961
Curium	Cm	96	(247)
Dysprosium	Dy	66	162,50
Einsteinium	Es	99	(252)
Erbium	Er	68	167,26
Scandium	Sc	21	44,955910
Tin	Sn	50	118,710

Strontium	Sr	38	87,62
Europium	Eu	63	151,965
Fermium	Fm	100	(257)
Fluor	F	9	18,9984032
Phosphor	P	15	30,973762
Francium	Fr	87	(223)
Gadolinium	Gd	64	157,25
Galion	Ga	31	69,723
Germanium	Ge	32	72,61
Hafnium	Hf	72	178,49
Helium	He	2	4,002602
Hydrogen	H	1	1,00794
Iron	FE	26	55,847
Holmium	Ho	67	164,93032
Indium	In	49	114,82
Iridium	Ir	77	192,22
Ytterbium	Yb	70	173,04
Ytrium	Y	39	88,90585
Krypton	Kr	36	83,80
Lanthanum	La	57	138,9055
Lawrencium	Lr	103	(260)
Lithium	Li	3	6,941
Lutetium	Lu	71	174,967
Magnesium	Mg	12	24,3050
Manganese	Mn	25	54,93805
Mendelevium	Md	101	(258)
Mercury	Hg	80	200,59
Molybdenum	Mo	42	95,94
Neodymium	Nd	60	144,24
Neon	Ne	10	20,1797
Neptunium	Np	93	(237)
Niobium	Nb	41	92,90638
Níckel	Ni	28	58,69
Nitrogen	N	7	14,00674
Nobelium	No	102	(259)
Gold	Au	79	196,96654
Osmium	Os	76	190,2
Oxygen	O	8	15,9994
Palladium	Pd	46	106,42

Silver	Ag	47	107,8682
Platinum	Pt	78	159,08
Lead	Pb	82	207,2
Plutonium	Pu	94	(244)
Polonium	Po	84	(209)
Potassium	K	19	39,0983
Praseodymium	Pr	59	140,90765
Promethium	Pm	61	(145)
Protactinium	Pa	91	231,03588
Radio	Ra	88	(226)
Radon	Rn	86	(222)
Rhenium	Re	75	186,207
Rhodium	Rh	45	102,90550
Rubidium	Rb	37	85,4678
Ruthenium	Ru	44	101,07
Samarium	Sm	62	150,36
Selenium	Se	34	78,96
Silicon	Si	14	28,0855
Sodium	Na	11	22,989768
Thallium	Tl	81	204,3833
Tantalum	Ta	73	180,9479
Tenacious	Tc	43	(98)
Tellurium	Te	52	127,60
Terbium	Tb	65	158,92534
Titanium	Ti	22	47,88
Thorium	Th	90	232,0381
Thulium	Tm	69	168,93421
Tungsten	W	74	183,85
Unnilcuadio	Unq	104	(261)
Unnilhexio	Unh	106	(263)
Unnilpentio	Unp	105	(262)
Unnilseptio	Uns	107	(262)
Uranium	U	92	238,0289
Vanadium	V	23	50,9415
Xenon	Xe	54	131,29
Iodine	I	53	126,90447
Zinc	Zn	30	65,39

Zirconium	Zr	40	91,224

* Atomic weights in parentheses are for the most stable isotope of the element.

A book as a travel, begins with concern and sadness ends.

José Vasconcelos.